# A2 in a Week

# Geography

Elizabeth Elam and
John Milner,
Abbey College, Manchester
Series Editor: Kevin Byrne

## Where to find the information you need

## SUCCESS OR YOUR MONEY BACK

Letts' market leading series A2 in a Week gives you everything you need for exam success. We're so confident that they're the best revision books you can buy that if you don't make the grade we will give you your money back!

### HERE'S HOW IT WORKS

**Register** the Letts A2 in a Week guide you buy by writing to us within 28 days of purchase with the following information:

- Name
- Address
- Postcode
- Subject of A2 in a Week book bought

**Please include your till receipt**

To make a **claim**, compare your results to the grades below. If any of your grades qualify for a refund, make a claim by writing to us within 28 days of getting your results, enclosing a copy of your original exam slip. If you do not register, you won't be able to make a claim after you receive your results.

### CLAIM IF...

You are an A2 (A Level) student and do not get grade E or above.
You are a Scottish Higher level student and do not get a grade C or above.
This offer is not open to Scottish students taking SCE Higher Grade, or Intermediate qualifications.

Letts Educational
Chiswick Centre
414 Chiswick High Road
London W4 5TF

Tel: 020 8996 3333
Fax: 020 8742 8390
e-mail: mail@lettsed.co.uk
website: www.letts-education.com

Registration and claim address:
Letts Success or Your Money Back Offer, Letts Educational, Chiswick Centre, 414 Chiswick High Road, London W4 5TF

### TERMS AND CONDITIONS

1. Applies to the Letts A2 in a Week series only
2. Registration of purchases must be received by Letts Educational within 28 days of the purchase date
3. Registration must be accompanied by a valid till receipt
4. All money back claims must be received by Letts Educational within 28 days of receiving exam results
5. All claims must be accompanied by a letter stating the claim and a copy of the relevant exam results slip
6. Claims will be invalid if they do not match with the original registered subjects
7. Letts Educational reserves the right to seek confirmation of the level of entry of the claimant
8. Responsibility cannot be accepted for lost, delayed or damaged applications, or applications received outside of the stated registration/claim timescales
9. Proof of posting will not be accepted as proof of delivery
10. Offer only available to A2 students studying within the UK
11. SUCCESS OR YOUR MONEY BACK is promoted by Letts Educational, Chiswick Centre, 414 Chiswick High Road, London W4 5TF
12. Registration indicates a complete acceptance of these rules
13. Illegible entries will be disqualified
14. In all matters, the decision of Letts Educational will be final and no correspondence will be entered into

Every effort has been made to trace copyright holders and obtain their permission for the use of copyright material. The authors and publishers will gladly receive information enabling them to rectify any error or omission in subsequent editions.

First published 2001
Reprinted 2002
Reprinted 2004

Text © Elizabeth Elam and John Milner 2001
Design and illustration © Letts Educational Ltd 2001

**British Library Cataloguing in Publication Data**
A CIP record for this book is available from the British Library.

ISBN 1 84315 366 1

Cover design by Purple, London

Prepared by *specialist* publishing services, Milton Keynes

Printed in the UK

Letts Educational Limited is a division of Granada Learning Limited, part of Granada plc

# Plate Tectonics and Hazards

**20 minutes**

## Test your knowledge

1. State the differences between continental and oceanic crust.

2. Earthquakes are caused by a build-up of _____ within the crust. Its sudden release creates _____ waves concentrated on a surface _____ and on a subsurface _____ .

3. How do the secondary effects of earthquakes contribute to a high death toll?

4. Why are the authorities very careful when issuing earthquake and volcano warnings?

5. Why might buildings be put on 'stilts' in areas subject to volcanoes?

6. Why is it difficult to issue emergency warnings and evacuation orders in LEDCs?

**Answers**

1 Continental crust is thicker (up to 75km), and less dense. It is older and comprised of mainly silica and basalt. Oceanic crust is only 6–11 km thick, more dense and made of mostly basalt. 2 pressure, seismic, epicentre, focus 3 Fires, floods, water-related diseases and famine may all occur after the quake. Fires are often fuelled by broken gas pipes and cannot be extinguished when water mains have burst. 4 Specialists are reluctant to give too many 'false alarms', as people may start to ignore the warnings. 5 Stilts would be useless against lava, but may protect against lahars. 6 Few people have television, radio or telephones.

✔ **If you got them all right, skip to page 10**

# Plate Tectonics and Hazards

**45 minutes**

# Improve your knowledge

1. **Plate tectonic theory** developed from Wegener's theory of continental drift. The evidence for this is:

   - the 'jigsaw fit' of the continents

   - the fossils of mesosaurus which were found only in southern Africa and Brazil

   - the geological match between rocks in North America and Europe.

   In 1948, the Mid-Atlantic Ridge was discovered, a continuous mountain range linking Iceland in the north with Tristan da Cunha Island in the south. It was found that rock in this ridge was young compared to that at the continental margins and that the Atlantic was widening by approximately 5 cm a year due to this 'sea floor spreading'.

   The parallel bands of rock either side of the ridge revealed that iron mineral particles aligned themselves to the pole as the lava cooled. As these rocks were dated, it was discovered that the regular reversals of the Earth's magnetic field were shown in these alternate rock bands and that these are almost symmetrical either side of the ridge (palaeomagnetism).

   The Mid-Atlantic Ridge is an example of a constructive or divergent plate boundary, where two oceanic plates move apart. Magma constantly rises up to fill the opening, building up into an oceanic ridge. The peaks of such ridges are submarine volcanoes and may grow above sea level (e.g. Surtsey near Iceland).

   A Constructive Plate Boundary

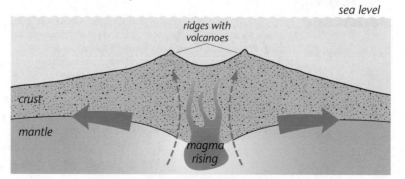

**Key points from AS in a Week**

| | |
|---|---|
| Floods | page 9 |
| Volcano Types | page 27 |
| Landslides | page 30 |

*Have a look at this 'jigsaw fit' or a world map in an atlas or a textbook.*

*Study a map of the plate boundaries in your textbook. Learn two or three of their main features.*

# Plate Tectonics and Hazards

At destructive boundaries, denser oceanic crust is forced under continental crust (e.g. along the Pacific coast of South America). Here, a long oceanic trench is formed in the subduction zone and the subducted oceanic crust melts due to the heat from intense friction. This magma forces its way to the surface to form volcanoes and fold mountains on land (e.g. the Andes) or island arcs at sea (e.g. off Japan).

As well as causing danger from lava, volcanoes also eject poisonous gases, ash and rocks and can cause floods, pyroclastic clouds (clouds of hot gas, molten rock and ash) and huge mudflows (lahars). Often, the lava is slow-moving. In this case, people have plenty of time to evacuate to escape the hazard, but cannot stop the lava destroying their homes.

At conservative plate margins, two plates slide past each other and crust is neither created nor destroyed. The two plates move horizontally alongside each other along a huge transform fault (e.g. the San Andreas fault in California). Violent earthquakes result from the sudden jolt or slippage.

*Draw simple labelled diagrams for the destructive and conservative margins as well.*

**2** Earthquakes are caused by a build-up of pressure within the crust. Its sudden release creates shock waves or seismic waves, which are concentrated on a surface epicentre and on a subsurface focus. Buildings collapse, cracks in the ground open up, roads and railways may be damaged and water and electricity may be cut off.

Earthquakes at sea can cause tsunamis or giant waves, which drown thousands and destroy harvests. They are generated by movements in the ocean floor, either directly due to an earthquake or due to slumping of undersea sediments near tectonic subduction trenches.

The high waves can travel thousands of kilometres before reaching land and flooding coastal areas. Small tsunamis caused by minor earthquakes can be seen quite frequently as unusually large 'freak' waves which travel further up the beach.

**3**  The plate margins are often heavily populated because:

- People perceive the risk as being minimal compared to the day-to-day problems of their lives.

- Many volcanic areas have very fertile soil or may contain valuable minerals such as gold and copper.

- Dormant volcanoes provide good defensive sites.

- People have the technology and money to cope with the hazard.

## Secondary effects

Earthquakes usually kill far more people through the resultant fires, floods, disease (especially water-related diseases) and starvation than by collapsing buildings. In the very long term, whole communities can disappear as people are afraid to return to the area and harvests are disrupted.

*Consider the different impacts of volcanoes and earthquakes in MEDCs and LEDCs.*

The damage the hazard causes depends on the level of risk. Risk is increased by:

- increased power of the earthquake/volcano/cyclone, etc.

- higher frequency of the hazard event

- people affected being of a low socio-economic status

- the level of economic development being low

- the hazard being in a rural or remote area

- low technology levels.

## 4   Forecasting and predicting hazards

Some hazards can be predicted, but this is more difficult to do with others. Certain regions, such as coastal Bangladesh, flood regularly and drought often occurs at five- or seven-year intervals at the same time of year. Earthquakes can be predicted to some extent by emissions of radon gas and minor tremors which may precede the mainshock. Animal behaviour is used to predict earthquakes. Unlike humans, animals seem to have an instinct that makes them behave in strange ways before a quake. Fish have been known to jump out of the water and dogs sometimes start howling before the earthquake begins.

Major known volcanoes are sometimes monitored by scientists to predict eruptions before they happen. They use sensitive instruments to measure any Earth tremors caused by magma moving, measure rises in pressure inside the volcano and analyse any gases released.

They also carefully research the volcano's history of previous eruptions to see if there is a pattern which might be repeated. Some volcanoes seem to erupt every 20 to 25 years, others every 100 to 120 years, and so on.

When warnings can be given, usually in more economically developed regions such as California, governments and industry can prepare and organise rescue efforts. Although some hazards can be predicted, some more accurately than others, specialists are reluctant to give too many 'false alarms', as people may ignore subsequent warnings.

**Modifying the hazard and damage limitation**

Some hazards can be modified to reduce the amount of destruction caused:

- Scientists have experimented with 'lubricating' the San Andreas Fault in California. They inject sand or liquids into the fault so that movement occurs without sudden jolts.

- The spread of forest fires can be stopped by burning a 500 metre-wide clear gap between sections of forest.

- When a volcano erupts regularly, such as Etna in Italy, the streams of lava often flow down the same channels. These streams can sometimes be diverted to flow in a different direction.

- Sometimes, large trenches are dug out of the ground. These carry the flowing lava away from settlements in case of an eruption.

- In Iceland, water is sprayed on the lava to harden it at one side so that the flow goes round nearby villages or towns.

- In many countries, frequent lahars or mudslides occur. Buildings can be put on stilts so that the mudflow runs underneath, causing less damage.

Evacuation is often the most important priority, but it may create more problems:

- If the hazard occurs near an international boundary, it is difficult for people to evacuate.

- Refugee camps are difficult to organise. After Mount Pinatubo erupted in the Philippines, more than 400 people died due to disease outbreaks in the camps.

- After the danger has passed it is difficult to move people back into the area with no services or infrastructure.

In many more economically developed countries, such as Japan, buildings can be reinforced with cross bracings, and flexible gas mains can be used to reduce the risk of fire. Some buildings have alarms that cut off the gas supply automatically and windows and furniture may be adapted to reduce injuries.

*Write a similar paragraph describing the measures that could be taken in a LEDC.*

To cope with ground liquefaction, buildings are constructed on floating concrete rafts.

## 6 Different responses

Countries differ in their responses to a hazard event. MEDCs have more resources to use for disaster relief, for fire control, for organising emergency procedures, for hospitals, for emergency water and food, for temporary accommodation such as tents and for special teams equipped to search for survivors. These teams may have infra-red heat-seeking devices to search for survivors in the rubble. LEDCs will have fewer, if any, of these facilities and sometimes valuable days can be lost whilst the army and police try to deal with the unexpected event. It is difficult to issue emergency warnings and evacuation orders as few people have television, radio or telephones. Buildings in LEDCs are more likely to be flimsy and not designed to withstand the earthquake.

Consider other hazards such as landslides, floods, cyclones, disease, drought, fires, pollution and crime. Think about:

- primary effects

- secondary effects

- human responses to the hazard.

# Plate Tectonics and Hazards

## Use your knowledge

1. (a) Why are earthquakes more likely than volcanic eruptions to occur as the main form of tectonic activity along some plate margins? Refer to one area along a plate margin that you have studied.

   (b) Using examples, discuss the ways in which people attempt to adjust to earthquakes as a hazard threat.

2. (a) Explain how volcanic activity can sometimes occur well away from the plate boundaries.

   (b) Describe how magnitude and frequency can affect the impact of different hazard types.

### Hint

Start by considering the different types of plate margin.

You could refer to differing approaches in 'developed' and 'developing' countries.

# Meteorology and Climate

**30 minutes**

 your knowledge

 Name the atmospheric system which transfers heat from the warm Equator to the colder North and South Poles?

 Name the three cells (Pole to Equator), in the three-cell model of atmospheric circulation.

 If an air mass forms over ocean it is called a _____ air mass and a _____ air mass if formed over land.

 Precipitation results from condensation of water vapour due to it rising and then cooling. Briefly describe the three ways in which this may be caused.

 Anticyclones contain subsiding (falling) air, which _____ and so holds more moisture. They are large areas of ____-pressure, bringing hot, clear days in summer and cold, foggy conditions in winter.

 What are the particles of dust called which help the formation of rain droplets and fog, and which cities create a great deal of?

 Give two other names for a cyclone.

 The Earth is heated during the day by insolation ( _____-wave radiation) from the sun. At night it is cooled by outgoing ____-wave infra-red radiation.

 What is the main source of sulphur oxides which cause acid rain?

✔ **If you got them all right, skip to page 20**

# Meteorology and Climate

**60 minutes**

## Improve your knowledge

**1** The Earth's temperature stays relatively constant. The insolation it receives and the outgoing radiation must be balanced. Although the insolation received varies with the length of night and day and the warm and cold seasons, the following factors are also important:

- Oceanic regions such as the Pacific can hold more heat energy than land surfaces and can store this heat in winter.

- Winds can blow from warm to colder areas, thus warming the latter up.

- Aspect is the direction in which a hillside faces. South-facing slopes are warmer than those facing north.

The deficit of heat at the North and South Poles and surplus in tropical regions is balanced by the heat budget. Otherwise, tropical regions near the Equator would overheat.

Heat is transferred by:

| Horizontal transfer | Vertical transfer |
| --- | --- |
| ocean currents | convection |
| 'jet stream' winds | conduction |
| cyclones/hurricanes | radiation |
| depressions | transfer of latent heat |

**2** In the atmospheric circulation model, the cells of circulating air create areas of convergence, with rising air and low pressure, or divergence, where air descends and pressure is high. These zones produce different climatic belts. (See illustration on next page.)

At the Equator, the Hadley cells of the northern and southern hemispheres meet. This creates a powerful uplift of warm air and frequent convectional thunder storms. This zone is called the Inter-Tropical Convergence Zone or ITCZ.

**Key points from AS in a Week**

Earth's Temperature
page 39

*You should be able to write a full paragraph to explain processes such as these.*

Atmospheric Circulation Model

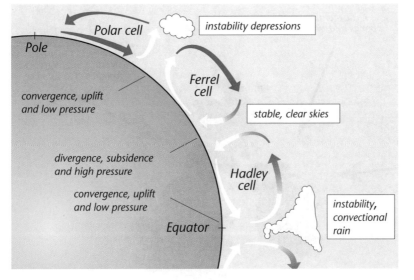

Atmospheric circulation is also affected by high speed, high altitude winds called Rossby waves. These encircle the Earth and give alternately rising or falling warm or cold air. They are wave-like bands travelling at up to 200 km/hr and are often used by balloonists to travel long distances quickly.

Within the Rossby waves are narrow waves of even faster winds called jet streams. The Polar Front Jet Stream or PFJS occurs where the Polar and Ferrel cells meet and affects the movement of depressions and anticyclones over the British Isles. The Subtropical Jet Stream or STJS marks the boundary between the Ferrel and Hadley cells.

**3** Large air masses affect the climate of the British Isles. They mostly originate in either the sub-tropics (e.g. the Sahara) or the high latitudes (e.g. Canada), where the air is stable and stationary. If they form over ocean they are maritime air masses and if they form over land they are continental. A maritime air mass holds more moisture and so more potential rainfall.

They slowly move either north or south, bringing particular characteristics, and cause instability when they meet. An air mass has uniform temperature and humidity and doesn't easily mix with a different air mass, with different temperature and humidity characteristics.

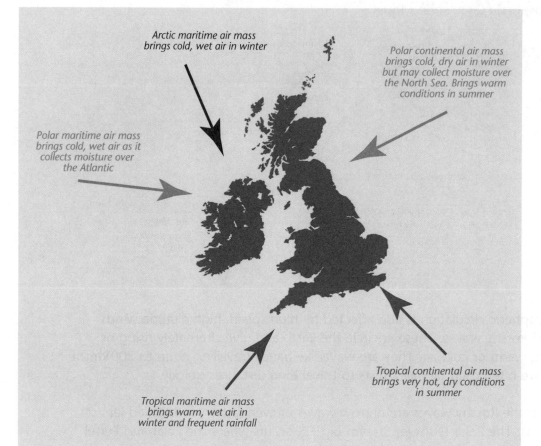

Arctic maritime air mass brings cold, wet air in winter

Polar continental air mass brings cold, dry air in winter but may collect moisture over the North Sea. Brings warm conditions in summer

Polar maritime air mass brings cold, wet air as it collects moisture over the Atlantic

Tropical continental air mass brings very hot, dry conditions in summer

Tropical maritime air mass brings warm, wet air in winter and frequent rainfall

The boundary between two air masses is called a front. As the British Isles is situated where different air masses often meet, there are more fronts created and so frontal rainfall is common here.

**4** Humidity is a measure of the amount of water held in the atmosphere. Absolute humidity is measured in grams of water vapour per $m^3$ of air. The air can hold more water at higher temperatures.

Relative humidity (RH) measures the amount of water in the air as a percentage of what would be held if the air was saturated (100%). On a very humid day, the RH may be 80% but a desert may only have 15%.

*Humidity is measured in a weather station using a wet-and-dry-bulb hygrometer. This has two thermometers, one of which is kept moist. The water evaporates, causing cooling. The difference between the two can be converted to relative humidity.*

As unsaturated air cools, it becomes saturated when it reaches the dew point. Further cooling causes condensation into either droplets or ice crystals.

Temperature decreases with altitude in the troposphere (the environmental lapse rate). A reduction in pressure also causes air temperature to fall, so when air rises, its temperature will fall due to a fall in pressure (the adiabatic lapse rate).

*In just the same way, warm air inside a room condenses to form tiny droplets on cold windows.*

Lapse rates

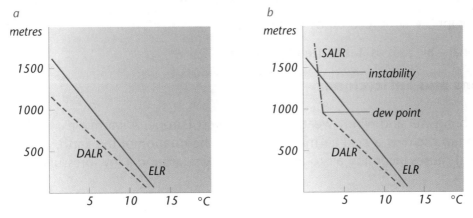

If the rising air is unsaturated, its temperature falls at a fixed, dry ALR of 9.8 °C per 1000 m.

If dew point is reached and condensation occurs because of this rising and cooling, the air may continue to rise, but whilst doing so it will cool at a slower and variable saturated ALR of between 4 °C and 9 °C per 1000 m.

Up to 1000 m, the air rising at the DALR is still cooler than the surrounding air (ELR) and so is stable (see diagram *a* above).

In diagram *b* above, dew point is reached at 1000 m and the air cools more slowly at the SALR. From 1500 m, the air is warmer than the surrounding air, which is cooling at the ELR, and so instability results in cumulus cloud and showers.

Droplets form in the cloud when water vapour condenses around dust particles. When the temperature is below zero, ice crystals will form, giving hail if they coalesce or snow if they cluster loosely in calm conditions. Rain is formed by the growth of ice crystals by sublimation, which melt as they fall (Bergeron-Findeisen process), or by the coalescence of droplets.

Therefore, precipitation results from condensation of water vapour due to rising and cooling. This necessary uplift is provided in three ways:

- cyclonic/frontal – warm air rises over cold in a depression

- relief/orographic – air forced up by mountains

- convectional – air rises as it is heated by the sun.

## 5 Depressions and anticyclones

Depressions bring wet and windy weather to northwest Europe. A warm air mass is forced to rise over a cold air mass at a front. The rotation of the Earth, (Coriolis force) causes incoming air to flow anti-clockwise around a low-pressure centre.

Anticyclones contain subsiding (or falling) air, which warms and so holds more moisture. They are large areas of high pressure bringing hot, clear days in summer and cold, foggy conditions in winter.

6 Large urban areas may have different annual patterns of temperature and precipitation. They have altered climates for the following reasons:

- Homes and industry generate heat.

- Homes and industry add more water to the atmosphere.

- Cities create dust (condensation nuclei), which helps the formation of rain droplets.

- Tall buildings alter winds and airflow.

- They have a different albedo or capacity to reflect heat.

# Meteorology and Climate

| Climate feature | Changes | Explanation |
| --- | --- | --- |
| Temperature | An 'urban heat island' exists whereby if winds are low, temperatures can be up to 4 or 5°C higher in the built-up city centre. | Buildings absorb heat during the day and release it slowly at night. Homes, industry and cars release heat. |
| Precipitation | Cities may receive up to 15% more rain and snow than surrounding rural areas.<br>Frequent occurrence of fog. | Urban heat can create strong thermals, which may cause heavy rainstorms.<br>Precipitation as fog is common due to abundance of dust particles. |
| Wind | Overall reduction in mean wind speeds but occasional 'funnelling' between tall buildings. | Houses and factories act as 'windbreaks' so reducing mean wind velocity. |
| Humidity | Lower relative humidity in cities. | Little vegetation so low evapotranspiration. Air can hold more water as it is warmer. |

*The falling temperatures towards the outskirts of an urban area are often shown by isotherms, concentric rings around the city.*

## 7  Climatic hazards

Cyclones are very deep depressions. Similar to the depressions which bring wet and windy weather to the British Isles, tropical cyclones have much more energy and can cause devastating storms, very high-speed winds and flooding. In America and the Caribbean they are called hurricanes, while in East Asia they are called typhoons.

They may be up to 2000 km across. In the Northern Hemisphere, the powerful winds flow anti-clockwise around the centre of the cyclone (clockwise in the Southern Hemisphere). They form over warm water and are most common at the end of summer when the water is warmest. This is around September in the Caribbean but November in the Indian Ocean.

In developed countries, warnings can be given days in advance and evacuations carried out, but in less economically developed countries, systems to cope with the hazard are often poorly developed and the level of devastation caused is far greater.

Features of a cyclone:

- high wind speeds up to 280 km/hr

- storm surges – huge waves which flood low-lying coastal areas

- torrential rain

- an 'eye' at the centre of the system – as the winds rotate around the eye, the centre of the storm is calm.

Effects of a cyclone:

- Flimsy buildings in LEDCs may be destroyed completely.

- Landslides often occur due to the heavy rainfall.

- Crops and harvests may be destroyed, causing local food shortages for several months.

- Damaged infrastructure can set the development process back many years.

Drought is an extended period of dry weather with little or no precipitation. Drought can cause famine due to the failure of crops. In desert or semi-desert areas such as the Sahel in North Africa, lower total amounts and reduced reliability of rainfall cause rivers and water holes to dry up, livestock dies and vegetation withers, the soil is exposed to the wind and desertification occurs.

In 1993 and 1994 much of Australia was affected by a prolonged drought. Most parts of the continent have long dry seasons – only the northern and eastern edges have more than 750 mm of precipitation per year. The plants and animals of Australia are adapted to live in drought conditions. However, in 1994 50% of sheep died and dry soil was blown away by the wind. Water restrictions were

*Learn how a recent cyclone affected the area over which it passed. Such natural disasters are always featured in newspapers. Include: the path of the cyclone; its effects/damage; how people coped.*

*Practise writing case study paragraphs such as this one for other hazards.*

imposed on people and dust storms were common in Australian cities. Numerous large bush fires also resulted from the drought, which reached the edge of residential areas, threatening lives.

**8** Global warming is caused by the greenhouse effect. The Earth is heated during the day by insolation (short-wave radiation) from the sun. At night it is cooled by outgoing long-wave infra-red radiation.

Around the Earth there is a blanket of 'greenhouse' gases, including carbon dioxide, methane and water vapour. These gases trap some of the outgoing radiation, keeping the Earth warm in the same way that a greenhouse heats up. Over the last 150 years there has been an increase in the amount of greenhouse gases in the atmosphere. This has mainly been due to the burning of fossil fuels such as coal and oil. As a result, more heat is being trapped, causing global warming. This will cause sea levels to rise (see *Coastal Environments*).

**9** Vehicles, industry and housing cause air pollution in cities. Photochemical smog is formed by a reaction between vehicle fumes and sunlight, especially in hot cities such as Los Angeles, where numerous laws exist to try to reduce car usage. Numbers of vehicles are increasing in most developing cities and now cities such as Bangkok have serious urban air pollution. Many pollutants can stay airborne over large distances – for example, sulphur oxides are carried from the UK to form acid rain, which harms lakes and forests in Sweden and Norway.

| Pollutant | Cause | Effects |
|---|---|---|
| carbon dioxide | power stations, cars and industry | causes the greenhouse effect and so global warming |
| carbon monoxide | cars, oil refineries and steel industry | poisonous; small amounts can damage brain functions |
| sulphur oxides | coal power stations | breathing problems, acid rain |
| lead | from vehicles' exhausts | brain damage |
| nitrogen oxide | from vehicles' exhausts | photochemical smog |

*You could write similar 'cause and effect' tables for other topics, such as inner city problems or migration.*

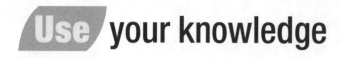

## Use your knowledge

**1** Study the map below.

London temperatures on a spring night

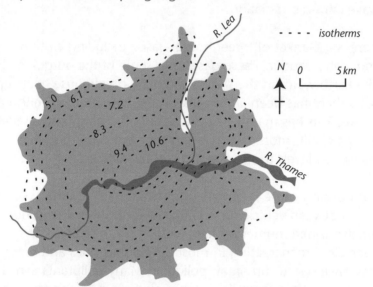

- - - - isotherms

R. Lea

R. Thames

N    0    5 km

*Even if you are presented with a simple diagram, study it carefully before you start to write your answer.*

(a) Describe how temperature varies across the city.

(b) Explain why temperatures seem to have this pattern.

(c) There is more total precipitation in urban areas than rural areas, yet rural areas receive more precipitation as snow. Explain this apparent contradiction.

*Don't mix up the 'describe' and 'explain' parts.*

**2** (a) Fully explain the causes of global warming.

(b) Describe the responses of governments, now that the threat of global warming is accepted.

*You need to refer to the environment summits.*

**20 minutes**

## Test your knowledge

**1** What is wave 'fetch' and why is it important?

**2** How do depressions contribute to the formation of storm surges?

**3** What are wave-cut platforms and wave-cut notches?

**4** What is the name for the ridge of larger pebbles sometimes found at the top of a beach?

**5** What are the two main types of sea level change?

**6** Why are some coastlines protected and others not?

## Answers

1 The distance of open water which the wind blows. Long fetch creates large waves. 2 Atmospheric low pressure allows sea level to rise up. High storm waves can then be blown up by the strong winds that are associated with depressions. 3 Wave-cut notches are formed when waves attack the base of the cliff and undercut it. Eventually, the overhang falls, leaving a straight cliff face and a wave-cut platform. 4 a storm beach 5 eustatic (worldwide change due to change in actual sea level) or isostatic (local change due to land-level change) 6 Many are left unprotected as they have low commercial value. Those that have residential settlement, valuable agricultural land or make income from tourism are protected by coastal defence measures.

**✓ If you got them all right, skip to page 26**

# Coastal Environments

## Improve your knowledge

 **Marine energy and waves**

Energy attacks and erodes coastal rocks and transports material. This energy comes from the development of waves. Waves are increased in size and energy by increases in wind velocity, wind duration and length of fetch (stretch of open water).

The dominant wave type (constructive or destructive) is that which has most effect on a certain coastline. Prevalent waves are those which affect the coast for most of the time.

Waves 'break' when the water depth is equal to their wave height. Swash flows up the beach and backwash down. A 'breaker' can be either plunging (destructive) or spilling (constructive).

Tidal range is the difference between high and low water mark. Tides can be classified as macro – where the range is over 4m, meso – 2 to 4m or micro – less than 2m.

 Storm surges are very fast rises in sea level. Seawater is pushed up against a stretch of coastline causing severe inland flooding. The consequences are even more disastrous when the coast is heavily populated. The southern part of the North Sea is particularly prone to storm surges when atmospheric depressions cause sea level to rise up, which may then be turned into high storm waves by the strong winds that are associated with such depressions. Processes of coastal erosion are more effective under storm conditions.

 **Features produced by coastal erosion**

- Cliffs: shape dependent on rock type and structure. Limestone and granite cliffs are usually steep due to slow erosion. Where rock is less resistant (such as shale), bays and headlands occur due to faster erosion. Waves refract either side of headlands because the water is shallower.

- arches, stacks and stumps
- caves and blowholes
- wave-cut platforms and wave-cut notches.

| Key points from AS in a Week | |
| --- | --- |
| Waves | page 12 |
| Tides | page 13 |
| Coastal Defences | page 17 |

*Many fieldwork techniques are straightforward to describe. The dimensions of waves (height and length) are just observed and then estimated using a long ruler. It is important always to take an average from many 'readings'.*

*Learn which specific erosional processes act upon the coast to form each feature.*

**4** Deposition occurs when the accumulation of material (sand and pebbles) is greater than its depletion, usually in sheltered areas with low-energy waves.

On shingle beaches, coarse material allows water to percolate through and so backwash is limited and shingle builds up on the beach. At the top of the beach a storm beach may form, made up of larger pebbles and boulders rolled up by the larger waves. On sandy beaches, the material does not allow water to pass through as easily and so there is considerable erosion by the backwash. Sandy beaches are likely to be less steep.

If waves break obliquely (at an angle) to the shore, the backwash is still at right angles to it and so material is transported along the beach (longshore drift).

*The relationship between beach particle size and beach gradient could be tested for fieldwork. Average particle size could be determined by random sampling and measuring.*

Longshore drift

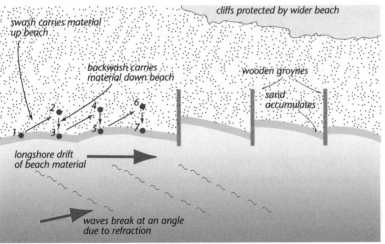

Spits form due to a loss of energy where a river enters the sea or where the coast changes direction. Material carried down the coastline by longshore drift builds up and is shaped by the direction of dominant waves.

Related features are: tombolos, bay bars, lagoons and cuspate forelands. Mudflats and salt marshes occur in sheltered water if deposition exceeds river erosion in river estuaries.

*You should be able to draw labelled sketches of each feature.*

**Sea level changes**

Changes in base sea level can be eustatic (worldwide change due to change in actual sea level) or isostatic (local change due to land-level change).

During ice ages, more water was stored at the poles and so there would have been an eustatic fall in sea level. At the same time, isostatic rises in sea level would have occurred due to the land 'sinking' under the weight of the ice. When the ice retreated, the land could rise due to the release of pressure, causing isostatic falls in sea level.

Tectonic processes can cause the land to uplift (orogeny) and hence lead to falling sea levels. Localised land tilting (epeirogeny) can cause coastlines to dip and be inundated by the sea.

Raised sea level creates submerged upland coasts with rias and fjords, or submerged lowland coasts with broad, shallow estuaries and mudflats.

Fjords were formed by rising sea levels, often at the end of the last ice age. They are glaciated 'u-shaped' valleys in mountainous areas and run perpendicular to the coast. When sea levels rose, these coastlines became submerged as the valleys were flooded.

Lowered sea level creates emerged upland coasts with a steeply rising shore and raised beaches, or emerged lowland coasts with coastal plains.

In the future, sea levels are expected to rise due to the continuing impact of global warming:

- melting mountain glaciers releasing stored water

- oceanic thermal expansion due to higher temperatures

- some melting of polar ice caps, although this will be offset by higher polar snowfall.

Although sea levels may only rise by up to one metre, large areas of low-lying coastal land may be inundated. Such areas are often agriculturally productive and highly populated. Many major world cities, including London, Calcutta and Tokyo, would be affected.

*Make sure you are clear about the way that uplifting land can cause falling sea levels and vice versa.*

*Look at the west coast of Norway or the southwest coast of New Zealand's South Island in an atlas.*

*The best way to revise features such as these is to draw a simple labelled diagram for each.*

# Coastal Environments

**6** ## Human interaction with coastlines

The human impact on coasts is the subject of significant debate and controversy. An example is the disruption of longshore drift with use of groynes.

Many coastal settlements are threatened by cliff retreat, and sea walls, revetments and breakwaters are used to slow the process. Coasts are also used extensively for recreation and the management of this must be considered.

Many coastlines are left unprotected as they have little commercial value. Coastlines that have extensive residential settlement, have valuable agricultural land or create income from tourism are usually protected by coastal defence measures. For example, many millions of pounds may have to be spent to protect the famous golf course at St Andrews on the east coast of Scotland, as the area's income is derived from golf tourism.

Coastal defence measures fall into two categories:

- 'hard' solutions involving building/engineering schemes

- 'soft' solutions which supplement natural beach protection.

**7** Most coastlines have significant pressures from different sources. These may include:

- Visiting tourists and day-trippers: coasts within easy reach of large population centres receive most visitors. This pressure is highest in the summer and during school holidays. Car parks and toilet facilities are necessary but may spoil the area's natural beauty. Footpaths to and from the beach become eroded.

- Agriculture: overstocking with cattle and sheep can cause dune areas to be stripped of vegetation and the sand to be blown away. Agricultural grasses can take over dune areas, displacing the natural flora.

- Sand and shingle extraction: many beaches have beach material extracted for use in the construction and road building industries. This lowers the beach level, bringing the sea closer, and removes the source of sand which is necessary to maintain the sand dunes.

*Governments and local authorities have to weigh up the cost of sea defences against the 'value' of the coastline in a particular area. How this 'value' is worked out is controversial.*

*Pressures and conflicts of interest may occur within each group. For example, windsurfers may come into conflict with anglers or bird watchers. The people involved will be different in each case study.*

*You will need detailed case study knowledge, including a sketch map, of the issues and possible solutions for a particular stretch of coastline.*

**45 minutes**

# Use your knowledge

 **1** Study the sketch map below.

Coastline sketch map

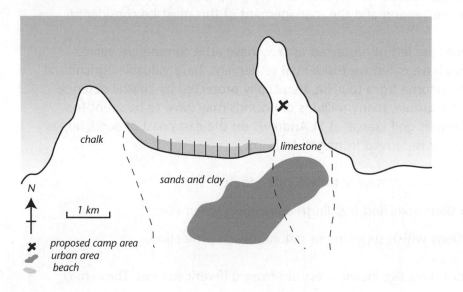

chalk

limestone

sands and clay

N

1 km

✗ proposed camp area
urban area
beach

*Use all the
information on
the map.*

(a) Explain how the geology of the area has influenced the shape of the
coastline.

(b) Explain what causes the groynes to be needed on the beach.

(c) Describe the possible environmental impact of the new campsite at X.

 **2** With reference to specific stretches of coastline, describe the various marine and
other processes which have been active, and the features produced.

*This question
directly asks for
case study
knowledge. Very
low marks given
for just describing
processes in
general.*

**20 minutes**

## Test your knowledge

**1** Within each ice age, fluctuations in the extent of the ice occur. Ice builds up at the poles and in mountainous areas and then retreats again. What are these periods called?

**2** In the short summers during an ice age, ablation exceeds accumulation and the mass and size of a glacier will:
(a) increase
(b) decrease
(c) stay the same.

**3** What is glacial abrasion?

**4** Why do corries form on the north slopes of mountains?

**5** Why are glacial meltwater streams able to entrain a lot of material such as sand and gravel as well as transporting larger rocks than would normally be possible?

## Answers

1 glacials and interglacials 2 decrease 3 Rock debris from freeze-thaw becomes embedded in the ice and scours the bedrock, either smoothing it or cutting lines or 'striations' on the rock surface. 4 'Aspect': north slopes receive less insolation from the sun and so the average temperature is lower than on south slopes. There is less ablation in summer or during the day so snow is able to build up. 5 During the summer, or at the end of a glacial period, the rate of ablation is rapid and so the meltwater streams have high discharge and velocity. Sometimes they flow in 'pipes' within the ice at high pressure.

✔ **If you got them all right, skip to page 32**

# Glacial Environments

## Improve your knowledge

In colder climates, either during previous ice ages or in mountainous polar regions, precipitation falls more often as snow. Cool, short summers mean that less ice and snow will melt. Snow becomes compressed by the weight of subsequent snowfalls on top, and so changes to a dense snow called firn. As more air is slowly pressed out, firn gradually changes into ice.

Key points from AS in a Week
Glacial Features
page 35

### 1 Evidence of glaciation

Scientists use ice-cores to obtain detailed climate data on periods up to 300,000 years ago. Cores over 3 km deep have been drilled in Antarctica, showing that major ice ages occur regularly, about every 100,000 years. Within each ice age, fluctuations in the extent of the ice occur, called glacials and interglacials. During a glacial, the average temperature only falls by 10°C or so, but this is sufficient to cause the build-up of ice at the poles and in mountainous areas.

The extent of the ice sheets at the peak of the last glacial

*ice sheet*

*Glacial features will only be found north of the line. Fluvioglacial features occur to the south of the line in some places.*

Depositional features – mostly terminal moraine – will be left at the furthest extent of the ice sheets, the ice margin. There may be an outwash plain extending away with drumlins and eskers. There will be a transition from unstratified (ice-dumped) to stratified or layered, fluvioglacial deposits.

# Glacial Environments

**2** Glaciers are large tongue-shaped masses of ice, which slowly move down a valley due to gravity. Under pressure, ice at the base melts and so allows the glacier to move. There is usually a meltwater stream, which runs continually in summer from the foot of the glacier. The surface is often uneven and cut by large cracks or crevasses.

The glacial budget is the net balance between accumulation (snowfall added to the glacier) and ablation or melting. In the short summers of an ice age, ablation will exceed accumulation and the mass and size of the glacier will decrease, whereas the opposite is true in winter.

At the surface of the glacier, the ice melts at 0 °C, as we would expect. However, within the glacier and especially at its base, ice can melt at lower temperatures, perhaps –2 or –3 °C. This occurs due to intense pressure from the weight of ice above. The ice is said to have a lower pressure melting point. Glaciers in temperate regions would therefore have more melting of ice at the base and it is this 'lubrication' which allows the glacier to move. Consequently, temperate glaciers have faster rates of movement than polar glaciers.

Speed of movement is increased by:

- steep gradients
- high accumulation from plentiful snowfall
- small glaciers responding quickly to falling temperatures
- relatively high summer temperatures.

**3 Processes of glacial erosion**

- freeze-thaw or frost shattering: most effective in areas with a summer thaw
- plucking: ice pulls away fragments of rock as the glacier moves
- abrasion: debris from freeze-thaw scours the bedrock
- rotational movement: allows for very effective erosion, so deepening the valley.

*Making lists like this for your other revision notes can help you to learn them more effectively.*

# Glacial Environments

## 4 Glacial features

You should be familiar with the characteristics and processes of formation of the following erosional glacial landforms:

- corries

- arêtes and pyramidal peaks

- glacial troughs ('u-shaped' valleys)

- truncated spurs and hanging valleys

- roches moutonnées.

*Draw a simple labelled diagram and write a paragraph explaining the formation of each*

Moraine

Moraine is a landscape feature which is formed when the rock material carried by a glacier is deposited.

| Type of glacial moraine | Description |
|---|---|
| lateral moraine | along the side of the glacier; comes from frost shattering of valley sides |
| medial moraine | moraine along the centre of the valley |
| terminal moraine | material pushed in front of the glacier |

# Glacial Environments

## 5 Fluvioglacial processes and features

Most glaciers have one or more meltwater streams running from the base of the glacier's 'snout'. These streams are the main component of ablation. There are two types:

- supraglacial: those flowing on the surface of the glacier

- subglacial: those flowing under the glacier.

During the summer, the rate of ablation is rapid and so the streams have a high discharge and velocity. Sometimes they flow in 'pipes' within the ice at high pressure. As a result, they can pick up (entrain) a lot of material such as sand, silt and gravel, as well as transporting larger rocks and other glacial debris than would normally be possible.

You should be familiar with the characteristics and processes of formation of the following fluvioglacial landforms:

*Draw a simple labelled diagram and write a paragraph explaining the formation of each.*

- Outwash plains and varves: flat expanses of sand and gravel deposited by the meltwater streams. Varves are alternating bands of coarser gravel and finer sand found in lake-bed deposits, corresponding to seasonal changes in stream discharge.

- Kettleholes: large blocks of ice may be buried in rock and soil. When the ice block melts a deep hollow is left in the ground.

- Braided streams: streams acquire numerous separating and rejoining channels, divided by low banks of shingle.

- Kame terraces and eskers: long banks of gravel and sand formed from river deposits left within the ice walls of the glacier.

# Glacial Environments

**45 minutes**

## Use your knowledge

 **1** Study the graphs below.

Rates of flow in polar and temperate glaciers

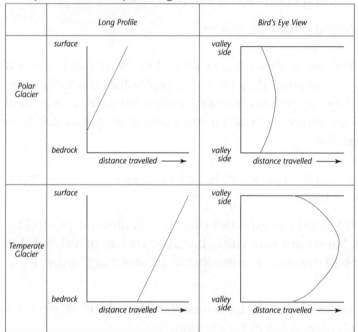

(a) Explain the different ways in which glaciers move.

(b) Describe the differences in the rates of flow between the two glacier types.

*No explanation yet!*

(c) Explain why the rates of movement are different.

**2** (a) Choose and name a landform of fluvioglacial deposition. Describe its typical characteristics and explain its formation.

(b) How can geographers distinguish between glacial depositional landforms produced by ice and those produced by water?

*Be careful! Many candidates mistakenly described an erosional feature.*

**20 minutes**

## Test your knowledge

**1** How does proximity to the sea affect the climate in periglacial areas?

**2** What is the 'active layer'?

**3** Why is chemical weathering not important in periglacial areas?

**4** What is 'segregation ice'?

**5** When do fluvial processes operate in periglacial areas?

**6** List four reasons why periglacial areas might be commercially exploited.

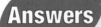

### Answers

**1** Periglacial areas near the sea have more variability – warmer winters with more precipitation. **2** The top layer, which may thaw each summer. Although partial thaw may occur near the surface in summer, the frozen ground below forms a layer impermeable to water. **3** Chemical weathering is not important due to the low temperatures and lack of available water. Even in summer when water is available, temperatures are still low, vastly reducing the rate of chemical weathering. **4** When water is drawn from the surrounding water-filled sediments into a large block or 'lens' of ice underground **5** when the temperatures are above freezing in summer but especially during the rapid spring thaw **6** military/strategic value, ports and airstrips, fossil fuels, metal ores, hydroelectric schemes, logging, fishing, fur trapping, or scientific research

**✔ If you got them all right, skip to page 38**

**30 minutes**

# Improve your knowledge

1  The term 'periglacial' is used to refer to areas peripheral to glaciers and ice sheets, which have similar geomorphologic and climatic characteristics. These areas are south of the polar regions (in the Northern Hemisphere) as far south as the treeline. The term can also apply to areas in cold mountainous regions that are not necessarily near to the poles.

Today, about 20% of the Earth's surface is periglacial, but during the last glacial period this would have been about 35%. Over 50% of both Canada and the Russian Federation are periglacial. The climate in these areas varies but common characteristics can be seen:

- Mean average temperature of 3°C or less. There are six to ten months when the temperature is below freezing and summer temperatures may peak at only 7°C.

- Temperature range may be very large, up to 50°C (–40°C in winter to +10°C in summer).

- Precipitation is low at only 200–300mm per year.

- Periglacial areas near the sea have more variability – warmer winters with more precipitation.

2  Periglacial areas have permafrost – the permanent freezing of the soil, sub-soil and bedrock up to 600m deep. Partial thaw may occur near the surface in summer, but the frozen ground below forms a layer impermeable to water. The top layer, which may thaw each summer, is called the active layer.

There are three types of permafrost:

- continuous: all of the ground is frozen, down to and including the bedrock

- discontinuous: most of the ground is frozen but thaw occurs in patches, called taliks

- sporadic: most of the ground thaws but a few frozen patches remain.

Factors affecting the depth and extent of permafrost

| Temperature | Negative heat balance is necessary. This occurs when terrestrial re-radiation exceeds insolation due to the low angle of incidence in high latitudes, and the short days. |
| --- | --- |
| Precipitation | If precipitation is reduced, permafrost increases in extent. |
| Aspect | Negative heat balance – more likely if it is north facing. |
| Albedo | High snow cover – much insolation is reflected back. |
| Ocean currents | The North Atlantic Drift reduces permafrost in northwest Europe. |

## 3 Processes and landforms of weathering

Chemical weathering is not important due to the low temperatures and lack of available water. Large angular blocks are produced by freeze-thaw action, (mechanical weathering) and on flat ground they do not move downslope but remain as blockfields or felsenmeer. There may be tors – large protrusions of weathered and jointed rock – and on slopes, talus cones of scree build up from weathered rock fragments.

## 4 Processes and landforms created by frozen ground

When the soil freezes downwards, stones can slowly be pulled to the surface due to frost heaving. If the ground has small domes (due to the expanding soil) then the stones will roll down to the dips between the domes, forming stone polygons up to 10m across. On slopes, elongated stripes occur as 'patterned ground' as the stones roll further downslope. Much larger stone polygons, up to 50m across, can be formed by large ice wedges in flat areas of freezing sand and silt.

Pingos are dome-shaped hills with a crater-like summit (see illustration overleaf).

*Your garden often has lots of stones on the surface after a cold winter.*

*You should practise writing detailed explanations of all the processes and resulting features.*

## Formation of pingos

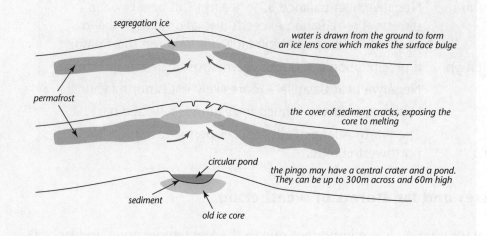

segregation ice

water is drawn from the ground to form
an ice lens core which makes the surface bulge

permafrost

the cover of sediment cracks, exposing the
core to melting

circular pond

the pingo may have a central crater and a pond.
They can be up to 300m across and 60m high

sediment

old ice core

 **5** Processes and landforms created by thaw

Annual river regimes are very variable: zero discharge in winter, then rapid
increase in discharge in spring, causing fluvial processes to operate. The large
rock fragments produced by freeze-thaw are entrained and transported by the
river as bedload. Braided channels are common (see page 31).

Spring meltwater cannot penetrate the permafrost so the active layer becomes
waterlogged. It may slip downslope due to gravity forming u-shaped
'solifluction lobes'. The rate of flow is increased by greater water content,
steeper slopes and sparse vegetation.

**6** Human activity in periglacial areas

Periglacial areas are hostile environments for people for the following reasons:

- no infrastructure (roads, fuel, water, etc.) – all has to be provided

- permafrost makes construction difficult

- taliks may appear under settlements

- long, cold and dark winters.

*Solifluction is a
'fast' process of
mass movement –
up to 2 m a day in
summer, but not
at all in winter
and perhaps only
a few cm a month
in spring and
autumn.*

# Periglacial Environments

Land use demands and conflicts exist between the local people and those who want to exploit the resources of such regions. The political power of those who support conservation is growing. They believe that rare natural ecosystems, unique landscapes and indigenous cultures should be protected.

Reasons why periglacial areas may be exploited:

- political: military or strategic value

- transport: air and sea routes, ports and airstrips

- non-renewable resources: fossil fuels, metal ores

- renewable resources: hydro-electric schemes, logging

- fishing and fur trapping

- scientific: research sites for meteorology, geology, biology.

You should study the ways that buildings and services can be adapted to periglacial areas and the consequences of human habitation on the permafrost structure and wider environment.

*Pressures and conflicts of interest may occur within each group. For example, fishermen may come into conflict with oil industry executives. The people involved will be different in each case study.*

*You will need detailed case study knowledge, including a sketch map, of the issues and possible solutions for a particular periglacial area.*

# Periglacial Environments

**45 minutes**

 **Use** your knowledge

**1**  (a)  Describe the typical climatic characteristics of periglacial areas.

(b)  Explain why the following features are all most active at certain times of the year:
   (i)   rock weathering
   (ii)  mass movements
   (iii) fluvial erosion, transportation and deposition.

(c)  Why do rivers often become 'braided' in such areas?

**2**  'It will always be difficult to settle permanently in periglacial areas.' Examine the reasons why this is true.

 **Hint**

*Quote average figures for temperature and precipitation. Show an awareness that there are different periglacial areas.*

*Consider the description you gave in question 1(a).*

*This type of essay question requires an introductory sentence and a concluding sentence.*

30 minutes

## Test your knowledge

 **1** Would the tundra regions be described as species-rich or species-poor?

 **2** What is a prisere?

 **3** Define the meaning of the word 'biome'.

 **4** Why do trees in the northern coniferous forests have a 'conical' shape?

 **5** Why do trees in the tropical grasslands (savanna) have small, waxy leaves and often shed leaves in the dry season?

 **6** How are the seeds of some desert plants adapted to the climate?

 **7** What are the four constituents of soil?

 **8** What is leaching?

**9** How does slopewise cultivation exacerbate soil erosion?

## Answers

**1** species-poor. **2** A prisere is a series of stages of colonisation by vegetation on a fresh or 'untenanted' site such as a lakeshore or exposed rock surface. **3** A biome is a climax vegetation type found in a large climatic region. **4** Conical shape gives stability against strong winds and winter storms and allows snow to be easily shed. **5** to reduce loss of water due to transpiration **6** They can lie dormant to wait for rainfall, then germinate and flower very quickly to reproduce again in a few weeks. **7** mineral matter, organic matter, air and water **8** Water dissolves calcium and other bases and washes it down through the soil. **9** The ploughed furrows become man-made gullies which allow rapid run-off.

✔ **If you got them all right, skip to page 47**

## Improve your knowledge

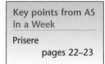

1. An ecosystem is an interacting community of flora and fauna together with the abiotic (not living) environment which includes the soil, climate and relief. Ecosystems vary in size from a puddle to a large climatic region. They will have various habitats, the niche occupied by a particular species.

The ecosystem may be:

- terrestrial (land based)
- aquatic (freshwater)
- marine (seawater).

Vegetation can be described in terms of its species composition and its layer structure. A plant community may be species-rich (e.g. tropical moist forest) or species-poor (e.g. tundra) and each species will have a different population.

All organisms require energy. For plants, this is derived from the sun. Other life forms derive their energy from eating plants or other animals.

In this way, a food chain develops, with plants as producers. Herbivores are primary consumers, carnivores are secondary consumers and the next level tertiary consumers.

Example of a woodland food chain

| Producers | Primary consumers | Secondary consumers | Tertiary consumers |
|-----------|-------------------|---------------------|--------------------|
| oak trees | caterpillars | sparrows | owls |

In addition, decomposers, including fungi and bacteria, break down dead matter to obtain energy for themselves.

Nutrients used by plants are stored in the biomass itself, in the soil or in the litter. The size of each store and the rate of flow of transfer between stores is

Key points from AS in a Week

Prisere
pages 22–23

*Easy marks for learning definitions!*

dependent on temperature and precipitation. In the boreal forest, the deep litter layer is the largest nutrient store, as low temperatures discourage humus incorporation into the soil by worms (low soil store) and biomass store is low as there is little undergrowth.

**2** A prisere is the series of stages of colonisation by vegetation on an untenanted site such as a fresh sand dune. The first plants to establish themselves form the pioneer community. With each stage, or sere, species diversity, biomass and soil fertility increase until the climatic climax community (CCC) exists, in equilibrium with the abiotic environment.

When an arresting factor (fire, deforestation, grazing) prevents the CCC from being reached, a plagio-climax exists with fewer species and lower biomass. If the arresting factor is removed, a secondary succession may result in the CCC.

There are four main types of prisere or succession with which you will need to be familiar: lithoseres (on bare rock); haloseres (on salt marshes); psammoseres (on sand dunes) and hydroseres (on lakeshores).

**3** A biome is a climax vegetation type found in a large climatic region. These generally run in latitudinal bands and include tundra, boreal forest, deciduous forest, temperate grasslands (prairie/steppe), tropical grasslands (savanna) and tropical moist forest. In mountainous areas, biomes change with altitude as well as latitude. Precipitation influences vegetation type (forest or grassland) and temperature determines rate of growth and whether the trees are deciduous or not.

You will need to be familiar with several biomes. The most popular are the tropical moist forest (or tropical rainforest), northern coniferous forest, tropical or temperate grasslands and arid/semi-arid areas. Questions may also be set on the vegetation of the British Isles. For each, you need to be aware of:

- geographical distribution
- climate: temperature extremes and annual range, precipitation total and annual distribution
- main species and their adaptations/response to the climate; vegetation layer structure (if any) and other chief features of the vegetation type

*These nutrient stores are often shown as three circles of different sizes. The relative values of the flows between them may be shown by arrows of variable thickness.*

*Very few areas have their true climatic climax vegetation due to interference by humans.*

*Study a world map of the different biomes in their latitudinal bands.*

*Make revision notes for the biomes you have studied using this list.*

# Ecosystems and Fragile Environments

- soil type and its features
- relative values of the three nutrient stores
- the effect of interference by man.

 **Northern coniferous forests**

| | |
|---|---|
| Location | High latitudes. Extends across North America, Northern Europe and Northern Asia. |
| Climate | Cold winters, perhaps −18°C. Short cool summers, 8°C. Low precipitation (350–650mm p.a.). In winter, frequent snowfalls. |
| Soil | Usually podzol, leaching can occur in higher rainfall areas. Deep, acidic, 'mor' humus layer. |
| Species | Few species. Spruce, firs and pines. Extensive, single-species stands. |

*Large-scale felling for paper and softwoods. Most countries have strict replanting programmes. Be prepared to write about a specific location, perhaps the strict control of logging in Scandinavia or Canada.*

Adaptations to the climate:

- Needle leaves reduce transpiration loss when water is unavailable (frozen) in winter.
- Evergreen foliage allows photosynthesis to begin immediately in early spring.
- Shallow roots mean that water can be obtained from the thawed surface, even if the subsoil is still frozen.
- Thick bark protects against cold.
- Conical shape gives stability against strong winds and allows snow to be easily shed.
- There is little undergrowth (only mosses and lichen) due to lack of light and frozen ground.
- There is a deep litter layer as there are few incorporators (worms) due to the cold.

 **Tropical grasslands**

| | |
|---|---|
| Location | Transitional belts between the tropical forests and semi-arid areas. Parts of Central, South and West Africa, central and western Australia, central Brazil. |

# Ecosystems and Fragile Environments

| | |
|---|---|
| Climate | Very hot summer (32°C), with some precipitation as intense convectional thunder storms. Long dry season in the 'warm' winter. Rainfall varies from 1200mm p.a. near forest edges to 250mm p.a. near desert edges. High evaporation loss. |
| Soil | Tropical red earths. Ferruginous soils – red colour from high iron oxide content. Soil is the largest nutrient store. |
| Species | Tall tufted grasses and occasional trees. Acacia and baobab trees. If rainfall is higher there are more trees and tall grasses ('tree savanna'). Only short, discontinuous grasses and bushes near desert edges ('shrub savanna'). |

Adaptations to the climate:

- Deep extensive roots ensure rapid growth in the rainy season.

- Withered 'straw' protects roots from sun in the dry season.

- Trees are short (6–12m), umbrella shape shields roots from the sun and streamlines against the strong trade winds.

- Small, waxy leaves and shedding leaves in dry season (deciduous) reduces transpiration loss.

- Some have broad, spongy trunk (e.g. baobab) to store water.

- Seeds can survive fires. Ashes fertilise new seedlings.

##  6 Arid and semi-arid areas

| | |
|---|---|
| Location | Two uneven 'bands', 15°–30° north and south of the equator. From the Sahara, across the Middle East into Asia, southwest USA, Kalahari in southwest Africa, Patagonia in South America, central Australia. |
| Climate | Hot deserts have high temperatures all year, 30°C in summer, 27°C in winter, clear skies with descending air. Low night temperatures. High diurnal range. Low precipitation. Negative water balance (deficit all year). High loss by evaporation. Generally less than 250mm p.a. Very irregular rainfall, occasional intense storms. |

*The African savanna is really a 'plagioclimax', the result of human activity: burning, over-grazing, tree felling. Nomadic herders are increasingly pressured to graze in just one region. Be prepared to write about a specific location, such as the Sahel.*

*Consider the effect of people in such regions. Desertification can occur due to over-cultivation, over-grazing, deforestation and salinisation due to irrigation. Be prepared to write about a specific location, e.g. the Sahel.*

| Soil | Saline soil (halomorphic) e.g. solonchak/solonetz. Often salt crust on surface. Weak, crumbly, calcium-rich. |
|---|---|
| Species | Drought-resistant and salt resistant (halophytic) thorny bushes (e.g. dwarf acacia), tufts of coarse, discontinuous grasses. Bulbous cacti. Few trees. |

Adaptations to the climate:

- Deep penetrating tap roots, spreading surface roots to collect moisture, large roots for storing water.
- Discontinuous cover – each plant can get enough water.
- Fleshy stems for storing water, tough bark to keep out heat.
- Leaves may be waxy/hairy/needle-shaped to reduce transpiration loss. Many have few or no leaves.
- Seeds can lie dormant to wait for rainfall, then germinate and flower very quickly.

## 7 Soil constituents and characteristics

There are four constituents of soil:

- mineral matter (derived from weathering of parent rock, influence recedes as soil ages)
- organic matter (comprises dead/decaying plant or animal matter and varied flora and fauna)
- air (rich in $CO_2$ and water vapour)
- soil water (varies in acidity/alkalinity).

Soil characteristics:

- Texture refers to the size of individual particles (clay, silt, sand, gravel) and determines water-holding capacity.
- Structure determines aeration and drainage and so affects fertility.

*'Open-textured' soils feel gritty and drain easily.*

Soil structure

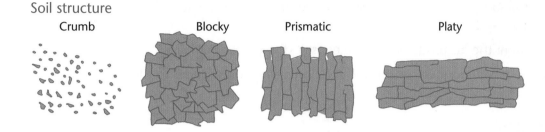

Crumb      Blocky      Prismatic      Platy

Nutrient recycling occurs when nutrients, absorbed through the roots of plants, are returned to the soil as leaf litter. Cation exchange is the process by which plants take up nutrients. Leaf litter is decomposed by earthworms or bacteria to form humus. This binds with clay particles weathered from the parent rock, forming the clay-humus complex.

Cation exchange capacity determines the fertility of the soil, and this in turn is related to availability of humus, suitability of conditions for soil fauna, presence of chemical compounds essential to plants, degree of acidity/alkalinity and texture/structure and drainage.

## 8 Soil formation and processes

- Climate: the seasonal and diurnal variations in temperature and precipitation affect the way soils form and develop.

- The type of parent rock influences the mineral content, texture and structure.

- The type and amount of organic life affects the supply of nutrients in litter and organisms such as earthworms, which incorporate them into the soil.

- Slope angle affects depth of soil and water flows.

- Chelation: nutrients and mineral ions are washed down the soil by organic compounds.

- Leaching: rainwater dissolves and washes down calcium and other bases.

- Eluviation: water washes down organic matter and minerals, undissolved.

- Illuviation: when a soil horizon/layer receives washed-down nutrients and minerals.

- Podzolisation: oxides of aluminium and iron are leached away from the upper horizons.

*Practise stating briefly the chemical changes which take place in each process.*

- Ferrallitisation: tropical soils can be red due to an accumulation of leached aluminium and iron oxides.
- Salinisation: the accumulation of excess salt due to low rainfall.
- Gleying: in waterlogged, anaerobic soils, red *ferric* iron reduces to blue *ferrous* iron.

 ## Soil erosion

This can occur in three ways:

- wind erosion when soil is *dry*, loosened by ploughing or bare after harvest/deforestation
- water erosion in sheets, rills or gullies
- chemical erosion or washing away of nutrients in high rainfall areas.

This natural process is exacerbated by human activity:

- Deforestation exposes soil to wind and rain.
- Overgrazing removes vegetation cover and exposes soil.
- Slopewise cultivation creates man-made gullies.
- Overcropping and monoculture exhaust the soil of particular nutrients.

Soil erosion can be slowed or even stopped by:

- afforestation to protect the soil
- tethering grazers (especially goats) to reduce overgrazing
- terracing and contour ploughing which slow run-off
- strip and cover cropping to reduce bare soil
- crop rotation, fallowing and adding fertilisers to reduce chemical erosion.

*You will be expected to refer to a case study of an area subjec to soil erosion. Try to evaluate the success of any measures taken to combat it.*

## Use your knowledge

**Hint**

*Your case study references must be specific and detailed.*

**1** (a) Changes to the climatic climax vegetation made by people have occurred to a greater extent in some parts of the world than others. Why is this the case?

(b) Describe how human activities have changed local ecosystems in semi-arid and arid areas.

**2** Study the diagram below which shows a representation of the nutrient cycle in an ecosystem:

Model of the nutrient cycle in an ecosystem

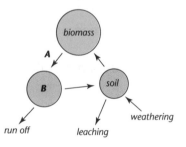

*Study the diagram for a full three minutes (use your watch!) before starting your answer.*

(a) Name flow A and store B.

(b) Explain how nutrients are transferred from store B to the soil.

(c) Explain how nutrients are transferred from the soil to the plants.

(d) How would the flows change if annual precipitation increased, perhaps as a consequence of global warming?

**15 minutes**

## Test your knowledge

 **1** How many people live in the world?

 **2** Why did population grow rapidly after the Industrial Revolution?

 **3** What rate of fertility is required for a population to replace itself?

 **4** What is the name given to the theoretical population that can obtain the highest standard of living in an area?

 **5** What are famines? Suggest a cause of famine.

 **6** What is food security?

 **7** Name three resources on which human populations depend.

 **8** Name the four most commonly used non-renewable energy resources in MEDCs.

 **9** What is OPEC?

**10** When might a resource like wood become non-renewable?

### Answers

1 approximately 6 billion 2 Accompanying improvements in sanitation, vaccines and standards of living raised life expectancy. 3 2.1 children per woman 4 optimum population 5 Famines are a chronic shortage of food in which many people die of starvation; causes include natural events (drought, natural disaster, crop failure, etc.), human mismanagement (land degradation, decline in access to food for those who most need it) and human inflicted problems (war, rise in prices). 6 access to sufficient food for all people in a country, not just enough food within the country 7 water, energy resources (non-renewable and renewable), soil, minerals 8 coal, oil, natural gas, uranium 9 Organisation of Petroleum Exporting Countries (influential in setting world oil prices) 10 if the rate of use does not match the rate of growth of biomass; in other words the use is unsustainable

✔ **If you got them all right, skip to page 55**

# Population and Resources

**45 minutes**

## Improve your knowledge

**1** The United Nations claimed that world population reached 6 billion in 1999. Between 1950 and 1990 the world population total doubled; the developing world total doubled between 1960 and 1980. Assuming that existing trends gathered from past data continue, population will continue to increase at a similar rate. However, in global terms there has been a decline in the rate of population growth from a peak of 2.1% in the late 1960s to approximately 1.3% in the 1990s.

**2** World population only grew slowly until the onslaught of the Industrial Revolution in Western Europe. In 1650, in the absence of any type of census, population was thought to be around half a billion. Population reached one billion around 1840, two billion in 1940 and 4 billion in the late 1970s. By 2040 it is predicted that it will have doubled again to eight billion. The huge and rapid increase since the 1930s is commonly known as the population explosion.

**3** The proportion and rate of population growth in different parts of the world is significant. Initially, population growth was high in the West, however, since 1960, most growth has been in LEDCs – in Africa, Asia and Latin America. Within this total, two countries, China and India, contain nearly a third of the world's population. Current average annual growth rates are 0.58% in MEDCs and 1.92% in LEDCs. According to the United Nations, 97% of the world's population growth is taking place in LEDCs. Although fertility has started to decrease in many LEDCs, particularly in Asia and Latin America, Africa's annual growth rate remains very high at 2.36%. The fall in fertility rates in LEDCs is mainly credited to China's one child policy. However, countries with large populations continue to add many people to the world population each year because of their sheer size, momentum and often youthful populations.

However, it is not just birth rates that contribute to high natural increase. Death rates have fallen in many LEDCs and life expectancy is increasing. Often the fall in death rate is not accompanied by the same relative fall in birth rate that the demographic transition model would suggest.

Despite huge growth rates in some parts of the world, many countries in

**Key points from AS in a Week**

Population growth (demographic transition)   page 47

Anti-natalist population policies
page 47

Urban issues in MEDCs
pages 58–59

Urban issues in LEDCs
pages 58–59

Rural development and change in MEDCs
pages 65–66

*Projected growth rates can vary depending on the data used and the reliability of that data.*

*Annual growth rate can be calculated by dividing the population at the start of one year by the population at the beginning of the next, and multiplying by 100 to get a percentage.*

MEDCs have total fertility rates (TFR) below the rate of 2.1 that is required for a country to replace its population (the replacement rate). Examples of countries include the UK – TFR 1.7, China – TFR 1.3 and Italy – TFR 1.2. These countries face problems of supporting and meeting the needs of an ageing population.

**4** The number of people which, when working with the highest available resources, will produce the highest per capita economic return is the theoretical optimum population. If an area has a larger population it is overpopulated, and if the area has a smaller population it is under-populated.

*The 'area' could b a nation state o region.*

Some countries, such as Bangladesh and Ethiopia, are considered overpopulated as they have insufficient food, minerals and energy resources to support their population at a reasonable standard of living. Other countries, such as Canada, are possibly under-populated because all the available resources are not exploited to their full potential. Although living standards are high in Canada, they could be even higher.

Malthus and Boserup's theories have been explained in relation to resources and carrying capacity in *AS in a Week*. A third approach to population growth was taken in the limits to growth model of the Club of Rome in 1974. By studying five factors that determine and limit growth (population, pollution, agricultural production, natural resources and industrial output) they argued that there is a fixed number of people that the Earth can successfully sustain. Their theory is classed as neo-Malthusian because they suggest that the world system will collapse in 2100 because of a lack of resources. Thus the time in which to act to even out resources and reduce population growth is very short. However, the Club of Rome suggested that trends could be altered to establish stability.

**5** Despite the pessimistic views of neo-Malthusians, there is now more than enough food to feed the world's population. Yet 20% of people still suffer from malnutrition (a calorie intake of less than 2200 per day) and some areas suffer from recurrent famine. There are many reasons for this:

- Poor farming methods – due to poverty, lack of knowledge, insufficient access to land, low government funding and poor infrastructure – can hinder the production and distribution of food.

- Colonial infrastructure was organised to ensure efficient administration and focused on the coastal areas, while the interiors of countries remained under-developed and neglected. This resulted in a dual economy (see the chapter on Development Issues) whereby the focus of investment is the urban sector (often a primate city) whilst rural areas have poor accessibility.

- Land degradation occurs because the carrying capacity has been exceeded where population growth outstrips food supplies. In the Sahel region of Africa, declining soil fertility in areas of low rainfall can limit food supply. This is worsened as a result of overgrazing, over-cropping, salinisation, deforestation and use of marginal land. An increasing number of natural and environmental disasters such as droughts and floods disrupt food supplies.

- Under colonial administrations, labour was diverted from subsistence agriculture to the plantations and mines. Some countries have become less self-sufficient in food production and increasingly dependent on food imports. There is concentration on cash crop export instead of domestic food supply. LEDCs remain politically tied to MEDCs (neo-colonialism), depend on exports of primary commodities, and are unable to develop their manufacturing base unless they receive investment from TNCs.

- Many LEDCs face enormous debt. The price of cash crop exports has not risen as fast as the price of manufactured imports. LEDCs have borrowed money and have had to divert money to service those loans. The International Monetary Fund (IMF) has implemented structural adjustment policies to make LEDC economies more market oriented so they can repay debts. As a result, money has been reallocated from welfare programmes, food subsidies and education.

- Civil war has brought about a breakdown in economies, disrupted food supplies and diverted money to military expenditure. Refugees have made food supply problems worse in recipient countries.

*Marginal land in terms of agriculture describes land which is not ideal for growing crops and may become degraded easily.*

**6** Food security can be increased by:

- controlling population growth (but this is a long-term measure)
- expanding food production by focusing on rural development and land reform
- liberalising world trade, so LEDCs receive fairer prices for primary commodities on the world market

- encouraging sustainable agricultural development to encourage soil water conservation via education

- encouraging political stability.

*Land reform is the redistributio of land to redres imbalances.*

**7** A resource is something that can be used to satisfy the needs of people. Resources can be classified as natural or human; renewable or non-renewable.

A classification of resources

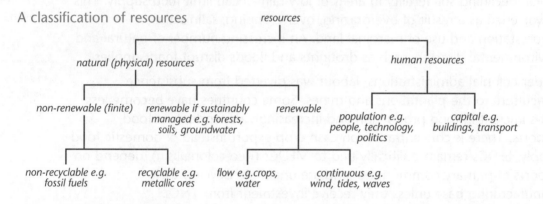

A stock describes the total amount of a given item. Stocks become resources only when there is a demand for them. Reserves are known resources that are considered exploitable under current economic and technological conditions. The strength of a country's resource base can greatly influence its level and speed of development. Most MEDCs have a wide range of resources and have the technology to exploit them. Many LEDCs are resource poor or their resources are controlled by TNCs.

*Less tangible resources includ attractive scenery, financia systems and research centres These are more diverse in locatio and are mainly associated with the service secto*

**8** Global resource use is very uneven – this is exemplified by energy resources, with 20% of the world's people using 70% of the world's fuels. Energy is the driving force behind most human activities, so it is fundamental to development. For the poorest people in the world, reliable and affordable energy supplies play an important role in improving living conditions, allowing them to use labour-saving devices and initiate small-scale industries.

Energy resources fall into two categories non-renewable (fossil fuels and nuclear power) and renewable (forces of nature such as water, geothermal and biomass). Fossil fuels provide 85% of the world's energy consumption; their distribution is very uneven, resulting in considerable world movement and trade.

Burning fossil fuels is one of the main contributors to greenhouse gases and the threat of global warming. Recent Earth Summits and climate talks in Rio, Kyoto and the Netherlands have attempted to draw up agreements to reduce consumption of fossil fuels and therefore emissions of greenhouse gases. However, whilst consumption has slowed very slightly due to recession in industrialised countries, LEDCs are aspiring to higher standards of living as their populations and industries grow.

*Most of the world's coal is located in Russia, North America and China. Most natural gas is in Russia and the Middle East. Most remaining oil is in the Middle East.*

**9** Factors affecting the type of fuel used within a country, exemplified by the UK:

| Factor | Situation in the UK |
| --- | --- |
| Availability, quality, lifetime and sustainability | The UK has always had abundant energy resources. |
| Cost of harnessing and transporting/importing | Exploitation of energy resources went hand in hand with industrialisation and rising incomes, so costs could be met. |
| Technology required to harness the resource | Improved technology allowed increased output per worker of coal and oil exploitation in the harsh North Sea. |
| Demands of the user (domestic, industrial, agricultural or transport) | Demand for coal has fallen for industrial use (steel) and domestic use. Power stations favour natural gas as it is cleaner and more efficient. |
| Political decisions | Nuclear power was promoted from the 1950s by successive governments as a solution to energy deficits. |
| Competition from other forms of energy | Natural gas and oil were discovered in the North Sea; this, along with competition from imported coal and high output costs, made coal uneconomic. |
| Environmental concerns | When fossil fuels are exhausted, British seas and weather have potential to provide renewable resources. Environmentalists argue that investment should be in renewables instead of nuclear power, which has great dangers associated with emissions and disposal of radioactive waste. |

*The UK has 300 years' worth of coal and 40 years of natural gas and oil.*

*OPEC has a major influence in determining world oil prices and production.*

# Population and Resources

**10** Sustainable development can be defined as development that meets the needs of the present without compromising the ability of future generations to meet their own needs.

The environment should be seen as an asset, a stock of available wealth. If the present generation spends this wealth without investment for the future then the world will run out of resources. If, however, we can research and develop new resources for the future, we can build machines that will substitute for the environmental resource. A good example is the construction of solar panels to replace oil and coal.

It is not just the unsustainable use of resources that is a pressing concern, but also the pollution of flow resources such as air and water. The sustainable yield is the amount that can be taken from a system without destroying it.

**40 minutes**

 **Use** your knowledge

 **Hint**

**1** (a) Explain what is meant by the term sustainable development.

    (b) What problems are likely to arise from the continued rapid growth of population In LEDCs?

    (c) How might sustainable development be achieved in LEDCs?

**2** What factors affect the rate of oil exploitation in the North Sea?

*Think of factors you can exemplify.*

*Think of economics as well as technology.*

**20 minutes**

## Test your knowledge

 **1** How does soil type affect agriculture?

 **2** What is an agribusiness?

 **3** Whose theory revolved around the idea of locational rent based on profit after transport costs?

 **4** What is the difference between nomadic pastoralists and shifting cultivators?

 **5** How can farms be viewed as systems?

 **6** Artificially watering crops is known as _____.

 **7** What are the advantages and disadvantages of removing hedgerows?

 **8** What is CAP?

 **9** What are three innovations of the 'green revolution'?

**10** What is eutrophication?

## Answers

**1** The ideal soil for farming is a loam; the texture allows good drainage and aeration and it has sufficient clay and humus to release nutrients. The pH is not too acidic. Soils can be improved via liming if they are too acidic, or by addition of fertilisers. Organic fertilisers add structure as well as nutrients. **2** Farming systems that are organised around scientific and business principles. They are linked to agricultural supply industries and food processing industries. They exhibit vertical integration and are part of complex agricultural chains, often linked to multinational companies. **3** Von-Thunen **4** Both move around – nomadic pastoralists raise animals whilst shifting cultivators grow crops. **5** They have input, processes and outputs, and positive and negative feedbacks. **6** irrigation **7** Fields are larger therefore there is more efficiency with machinery and a larger area under cultivation. Soil erosion is increased as there is less protection, cross pollinating insects are eliminated and monoculture is encouraged. **8** Common Agricultural Policy, the farming policy of the European Union **9** high-yielding varieties of rice, fertilisers, pesticides, irrigation schemes, mechanisation **10** nitrate pollution in fresh water supplies encouraging algal growth and oxygen demand which kills fish

✔ If you got them all right, skip to page 64

**45 minutes**

## Improve your knowledge

**1** A variety of factors affect the type of agricultural land use:

- Physical or environmental factors include: the length of the growing season and the number of frost-free days; the annual soil moisture budget and the balance between precipitation and evapo-transpiration; the wind chill factor; altitude, aspect and slope gradient and soil type. Global warming is probably a new factor that may lead to changes in temperatures and rainfall patterns.

*Soil moisture budget refers to water availability in different seasons.*

The optima and limits model

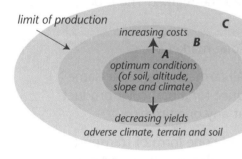

limit of production

increasing costs

C

B

A

optimum conditions
(of soil, altitude,
slope and climate)

decreasing yields
adverse climate, terrain and soil

**A** = OPTIMUM CONDITIONS Where the total cost of production per unit output is minimised for that crop or livestock.

**B** = DECREASING YIELDS as farming conditions become less than ideal. The profitability of the crop or livestock is reduced and the LAW OF DIMINISHING RETURNS operates when either the output decreases or the cost of maintaining high yields becomes prohibitive.

**C** = ADVERSE PHYSICAL CONDITIONS do not permit agricultural production on an economically viable scale or even at subsistence level.

- Economic factors include transport costs, market demand and consumer preferences, availability of capital, economies of scale, labour supplies and levels of technology/mechanisation.

- Social factors include land tenure (freehold, tenancy, share cropping), farm size (inheritance laws may fragment farm, whilst trends towards agribusiness has increased commercial farm size).

- Political factors include tariffs on imported food to protect home producers, subsidies to guarantee prices, grants to carry out improvements, and quotas to limit the amount produced.

**2** Farming may be classified in several ways:

- Arable farming is the cultivation of crops. Pastoral farming is the rearing of animals. Mixed farming includes both crops and animals.

- Intensive farming involves a high input and output per unit area of land. Extensive farming involves a low input and output per unit area of land.

- Intensive farming may be capital intensive or labour intensive.

- Nomadic farmers and shifting cultivators have no fixed abode; in contrast, sedentary farmers stay in the same place.

- Subsistence farmers produce food for themselves and their family and may trade any surplus or store it for lean years. Commercial farmers aim to make a profit by selling the product.

**3** Von Thunen's theory of agricultural land use was introduced in 1826. It is a concentric-ring model showing how agricultural land use varies with distance from the market. The most intensive farming is near the market and the farming becomes more extensive the greater the distance from the market. The model assumes:

- An isolated state with an isotropic plain and only one market.

- All farmers receive the same price for a particular product at any one time.

- Only one form of transport is available and transport costs are directly proportional to distance.

- Farmers act knowledgeably and aim to maximise profits.

*It is possible to apply Von Thunen's model at a variety of scales. On an international scale, the intensity of farming within the EU tends to decrease with distance from the core. On a local level, many present-day villages in Africa exhibit certain features of the model.*

*Isotropic plains are flat with no physical barriers.*

The relationship between locational rent and land use in Von-Thunen's model

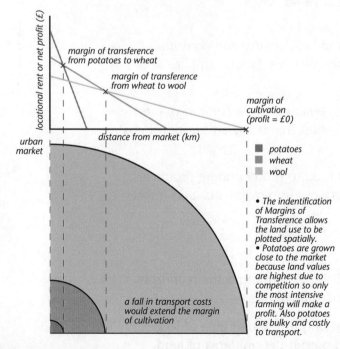

margin of transference from potatoes to wheat

margin of transference from wheat to wool

margin of cultivation (profit = £0)

locational rent or net profit (£)

distance from market (km)

urban market

- potatoes
- wheat
- wool

• The indentification of Margins of Transference allows the land use to be plotted spatially.
• Potatoes are grown close to the market because land values are highest due to competition so only the most intensive farming will make a profit. Also potatoes are bulky and costly to transport.

*a fall in transport costs would extend the margin of cultivation*

The model is based on the concept of locational rent or profit generated per unit area. This can be worked out as income received for a particular crop minus the production and transport costs. Due to the assumptions, the only factor that changes is transport costs and these are directly proportional to distance from the market.

# Rural Areas and Food Supply

 A comparison of farming systems to illustrate the factors affecting farming

## Farming system and factors affecting the farming system

### Intensive subsistence rice farming in Bangladesh

- Monsoon climate allows three main crops of rice.
- 75% of the population work in agriculture; any improvements in standards of living are dependent on increasing agricultural output; food production has not kept pace with population growth since 1971.
- Most land is concentrated in the hands of minority landlords; land reform is essential in order to redistribute land.
- Adequate and controlled flood water supplies are vital for the replenishment of the delta's fertility.

### Extensive commercial wheat farming on the Canadian Prairies

- Climate means growing season is short, precipitation is low and the chernozem soils, though fertile, are vulnerable to soil erosion.
- Relief is gently undulating, which aids machinery and transport.
- Input of capital is extremely high; farming is highly mechanised; large quantities of pesticides and fertilisers are used.
- Seed varieties have been improved to be disease-resistant, drought-resistant and faster growing.

### Nomadic herdsmen (Rendille) of Northern Kenya

- Rainfall is too low and unreliable to support sedentary agriculture, so people move with their belongings and animals to find pasture and watering holes.
- Land is becoming overpopulated and resources over-stretched as the number of people and animals increase and as water supplies and vegetation become scarcer; soil erosion is the result.
- People are forced to settle in towns and the use of firewood increases soil erosion leading to desertification in many areas where rainfall patterns are very low.

### Extensive commercial pastoralism in the Pampas, South America

- Favourable temperatures allow all-year grass growth.
- Many ranches exceed 100 square kilometres.
- Cattle have been bred to a give beef cows capable of living in warmer, drier conditions.
- The development of the rail network and refrigerated wagons allows export to industrialised countries.

# Rural Areas and Food Supply

 Farms can be viewed as systems with positive and negative feedback. For example, if outputs exceed inputs the equilibrium is maintained because capital is reinvested in the system (negative feedback). If the inputs are greater than the outputs, the farm will stagnate and decline as the equilibrium is lost (positive feedback, the runaway mechanism of change).

Irrigation is the provision of a supply of water from a river, lake or underground source to enable an area of land to be cultivated. It may be required where:

- rainfall is limited and evapo-transpiration exceeds precipitation (Nile delta)

- seasonal water shortages occur (Mediterranean)

- rainfall is unreliable (Sahel region of Africa)

- farming is intensive despite high levels of rainfall (rice-growing areas of Southeast Asia).

Water resources are often most lacking in LEDCs and these countries often lack the capital and technology to access water supplies. Many prestige schemes such as dam building have created debt and wide-ranging environmental problems, despite their regulating river regimes and providing flood control.

*The Aswan Dam on the River Nile is well documented.*

 Agriculture has undergone huge changes in the UK since 1945. The main features have been related to an intensification of agriculture:

- decrease in the number of farm workers by two thirds between 1944 and 1987 due to increased mechanisation

- increase in farm size and a decrease in the number of farms through amalgamation

- increase in the capital-intensive nature of farms

- increasing regional specialisation in terms of produce

- increasing corporate interest in farms to create agribusiness

- more recently, farmers diversifying their activities and an increase in organic farming.

# Rural Areas and Food Supply

**8** The Common Agricultural Policy (CAP) has had a profound impact upon farming in the EU countries. The objectives of CAP were to increase productivity, promote self-sufficiency and ensure reasonable prices and therefore standards of living for farmers.

The basic features of CAP were subsidies so EU farmers could compete on the world market and grants so farmers could improve their farms via mechanisation and land improvements. However, CAP was so successful that surpluses were produced. In the 1980s measures were taken to modify the policies to decrease overproduction. Also under consideration was the impact that intensification was having on the environment, such as loss of hedgerows, soil erosion, and loss of biodiversity.

*Surpluses included grain 'mountains', butter 'mountains' and wine 'lakes'.*

A scheme known as set aside was introduced, which paid farmers to take land out of production for at least five years. The land could remain fallow, be converted to woodland or be used for non-agricultural purposes (war games, camping, etc.).

*Biodiversity refers to the range of plant and animal species in an area.*

Quotas allow farmers to keep only a certain number of livestock and produce fixed amounts of beef, milk, etc.

Farmers in the EU now face issues concerned with animal health (such as BSE and foot and mouth disease) and genetically modified foods. Many people continue to leave the farming profession as price support has decreased dramatically since 1992.

**9** In LEDCs, the main changes in agriculture have been:

- the influence of the 'green revolution'
- the expansion of cash-crop acreage in response to debt
- increasing involvement of TNCs
- decline in rural population due to rapid urbanisation
- increased numbers of landless labourers
- increased mechanisation
- declining contribution of plantation agriculture and a shift to small-scale commercial farming

- less nomadism in favour of sedenterisation.

The green revolution refers to a package of measures introduced into LEDC agriculture from 1960 onwards. The aim was to expand food production to keep pace with rapid population growth:

- High Yielding Varieties (HYVs) were introduced.

- More fertilisers, irrigation and pest control were used.

- Mechanisation was increased.

*Sedenterisation means settling into one place.*

*An example of an HYV is 'miracle rice' which produced eight times as much rice per plant.*

| Positive impacts of the green revolution | Negative impacts of the green revolution |
|---|---|
| • Increased output especially in Asia; the world rice harvest doubled between 1967 and 1992. | • Environment has been damaged by too many fertilisers, and excessive irrigation has led to salinisation. |
| • Increased yields have led to a fall in food prices. | • HYVs are more susceptible to drought, pests and diseases (irrigated land is required). |
| • Faster-growing varieties allow an extra crop to be produced each year. | • HYVs are less nutritious and people do not find them as tasty. |
| • Allows other crops to be grown to add variety to the local diet. | • The already dominant local elite have gained even more power because they can afford to buy the seeds and fertilisers, etc. |
| • Many of the larger land-owners have become more wealthy. | |
| • HYVs also allow the production of some commercial crops. | • Increased landless labour due to mechanisation encouraging rural to urban migration. |
| • HYVs are not as tall, allowing them to withstand wind and rain. | |

The environmental impacts of farming include:

- loss of wildlife habitats such as wetlands and hedgerows

- accelerated soil erosion due to poor farming practices; rates of topsoil loss are very high in East Anglia, for example

- leaching of chemical fertilisers and pesticides into rivers and groundwater supplies leading to eutrophication in rivers (rapid growth of algae starves fish of oxygen)

- salinisation where irrigation waters are not properly drained away and a salt crust forms on the surface, rendering the land unproductive.

*Nearly half of the irrigated land in Egypt and Pakistan is affected by salinisation.*

**40 minutes**

## Use your knowledge

**1** How has the intensification of food supply systems changed ecosystems in MEDCs?

*Consider impacts on the land and knock-on effects.*

**2** What are the advantages and disadvantages of relying on a single cash crop?

*Think also about the value.*

**3** Discuss the idea that large farmers have been the only winners in the green revolution.

*Give a balanced view.*

# Manufacturing Industry

**15 minutes**

## Test your knowledge

**1** What is the term for the four categories of industry?

**2** In the Weber model of industrial location, the lines of equal transport costs are known as _____.

**3** What is GATT?

**4** What is structural change?

**5** What type of economy does the UK have?

**6** Give an example of a product that is now obsolete due to new technology.

**7** How can a TNC influence political decisions in a country?

reduction in taxes

jobs if the government does not allow location in a certain area or a
word processors and computers **7** because it may withdraw causing a loss of
privatised over the last 20 years **6** e.g. typewriters have been replaced by
tertiary, quaternary) **5** mixed, although many state industries have been
percentages of workers in each sector of employment (primary, secondary,
reduce barriers to world trade via 'rounds' of talks **4** change in the
**1** sectors **2** isotims **3** General Agreement on Tariffs and Trade; aims to

**✓ If you got them all right, skip to page 72**

# Manufacturing Industry

**45 minutes**

## Improve your knowledge

**1** Manufacturing (secondary) industry consists of companies that convert raw materials into finished goods or assemble components made by other manufacturing companies.

- Heavy industry requires bulky raw materials and large amounts of energy (e.g. ship building, iron and steel).

- Light industry uses small amounts of material per worker; it is not generally a source of pollution and is found on industrial estates.

- Hi-tech industry often involves using micro-electronics.

**2** Many geographers have devised models of industrial location to predict and explain where firms choose to locate. The summary on the next page explains some of the most important models and should be studied with the important concepts outlined below.

Agglomeration is when several firms choose the same area for their location in order to minimise costs. This can be achieved by linkages between firms and their supporting services. Firms are also able to achieve economies of scale with such concentrations. Agglomeration was one of the factors that Weber suggested could affect the location of industry.

Linkages are relationships between one industry and another:

- forward: to the consumer of the industry's product

- backward: to the provider of raw materials and components

- vertical: where the raw material goes through several successive processes

- horizontal: where an industry relies on several other industries to provide its component parts

- diagonal: where an industry makes a component which can be used in several subsequent industries

- technological: when a product from one industry is used as a raw material by a number of subsequent industries which further reprocess it.

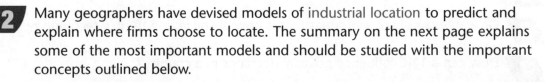

> **Key points from AS in a Week**
>
> Definitions of industry    page 78
>
> Factors affecting location of present-day industry
>                 page 79

## Theories of industrial location

### 1 Weber's model of industrial location

*Four factors affect production costs:*
* *cost of raw materials*
* *cost of transporting raw materials and finished product*
* *labour costs*
* *advantages or disadvantages of agglomeration (i.e. locating near other firms to create linkages which minimise costs).*

*The least cost location is identified by drawing isotims (lines joining all places with equal transport costs for either moving the raw material or the product) and then constructing isodapanes which join all places with equal TOTAL TRANSPORT COSTS. Critical isodapanes show the points at which savings made by reduced labour costs equal the losses brought about by extra transport costs. Moving away from the least cost location is profitable if cheaper labour lies within the critical isodapane.*

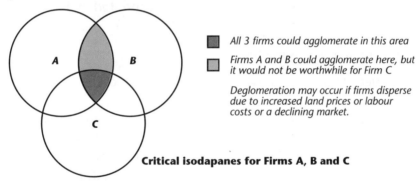

■ All 3 firms could agglomerate in this area

□ Firms A and B could agglomerate here, but it would not be worthwhile for Firm C

Deglomeration may occur if firms disperse due to increased land prices or labour costs or a declining market.

**Critical isodapanes for Firms A, B and C**

### 2 Smith's area of maximum profit

*Firms rarely locate at the least-cost location because they have imperfect knowledge of production costs and market demand. Firms may be encouraged to locate in particular areas e.g. of high unemployment.*

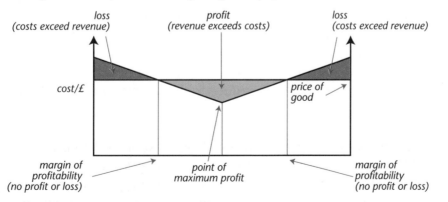

# Manufacturing Industry

Economies of scale are the factors that cause average costs per unit to be lower in large-scale operations than in small-scale ones.

Internal economies of scale include:

- specialisation: a large workforce has skills which match up exactly with job requirements
- bulk buying
- fixed costs of equipment are spread over more units of output.

External economies of scale operate when firms are concentrated in a geographical area:

- a specialised pool of labour is attracted
- networks of suppliers are attracted, which helps to reduce costs.

Footloose industries are those that have a relatively free choice of location and are not tied to the location of raw materials. Many are new industries that produce light goods of high value or a service. They are often located on new industrial estates on the edges of urban areas or along major motorways to utilise the efficient road transport system.

*The location quotient is a statistical measure of the degree of concentration of an industrial activity in a region compared with the national average.*

Industrial inertia occurs when an industry remains concentrated in an area even though the original factors that caused it to be located there no longer apply.

**3** Comparative advantage is the principle that countries can benefit from specialising in the production of goods at which they are relatively more efficient or skilled. In this way, the consumers within each country gain the maximum benefit from international trade. GATT accepts the notion that more trade benefits everyone and has sought to reduce trade barriers by agreements with major trading nations.

*Britain could produce cheap steel with available iron ore and coal; that comparative advantage is now lost.*

Free ports, such as those in Singapore, allow import and export of goods without taxation. This encourages manufacturing and assembly industries to locate in these countries. Import substitution is when a country establishes industry in its own country which can produce goods that were once imported.

NICs have pursued this policy and this has contributed to their success in showing rapid growth in manufacturing since the 1960s.

**4** Economic growth occurs when real incomes per head increase over time as a result of the more efficient use of the factors of production to provide more goods and services. Factors of production include land, labour, capital and enterprise.

*Diversification is the spreading of business risks by reducing dependence on one product or market; it enables companies with saturated markets to find new growth opportunities.*

Full employment provides jobs for everyone who wishes to work. What constitutes full employment is debatable because structural change seems to create higher levels of unemployment at every stage of the trade cycle. The regular pattern of upturns and downturns in the economy tend to repeat every five years or so, leading to booms and recessions, but a general trend of increased economic growth.

The Kondratieff cycle is a theory that there exists a 50-year cycle of economic upturn and downturn. The most widely accepted explanation for the cycle is that the introduction of new technologies causes disruption, but once the technology is established, it forms the basis for many new products and jobs. In the 1930s the car was replacing rail, and in the 1980s the microchip was replacing mechanical technology.

**5** Interventionists believe that government action is necessary to push an economy out of recession and that the free market will not bring about full employment. The 'laissez-faire' approach holds that the government should avoid intervention and that the free market will maximise business efficiency and customer satisfaction.

- Market economies allow markets to determine the allocation of resources.
- Command economies are entirely run by the government using central planning.
- Mixed economies are a compromise between the two (the UK has a mixed economy with some state-run industries and many privately-run industries).

Structural unemployment occurs when there is a change in demand or technology that causes long-term unemployment. Very often this occurs in particular regions which have been heavily dependent on certain industries, such as coal mining in south Wales and ship building in northeast England. Downward spirals may occur if a region loses the more motivated people and therefore less investment is attracted, which in turn means more people will out-migrate, resulting in even less investment. There is little that governments can do other than offer retraining for those made redundant or implement regional development schemes:

- Enterprise zones are areas set up by the government to attract industries by the removal of certain taxes and local authority control.

- The European Union gives regional assistance to many areas that have suffered industrial decline.

- Growth poles have been adopted by many governments to reduce regional imbalance. New economic development is targeted within an area of a country to attract inward investment and reduce out-migration. Often branch plants are located in these areas and are the first to close during times of economic recession.

- In the UK, various strategies have been attempted to regenerate both industry and declining urban areas, including assisted areas, development areas, enterprise zones and urban development corporations.

**6** Globalisation is increasingly important; it refers to the large-scale operation of physical, human and economic systems. Each country's economy has become less self-contained and more part of the global process of change.

Global shift refers to the locational movement of manufacturing production on the global scale. In the late 19th century and for much of the 20th century, manufacturing activity was concentrated in North America and Western Europe, but in the 1970s and 1980s TNCs (transnational companies) began to establish plants in the NICs. The traditional areas became more oriented towards tertiary activity and research and development. The focal point of manufacturing activity shifted to countries around the Pacific Rim.

The 'new industrial revolution' refers to a change in industry from Fordism (mass produced single items by a largely unionised workforce) to a flexible system of manufacturing with a more varied range of products within a

*Global growth of manufacturing output is now greater than the growth of employment; this has produced the 1990s concept of 'jobless growth'.*

*You will need to know case studies of regional development, such as the Mezzogiorno in Italy or south Wales.*

*The 'new international division of labour' refers to the split of skilled and routine assembly jobs between MEDCs and LEDCs.*

company which can respond quickly to market needs. Obsolescence occurs when a product, service or machine has been overtaken by a new idea that provides the same function in a better or more attractive way. This may result from new technology, but also a demand on the part of the customer who wishes to replace items that are in fact still usable.

**7** TNCs or MNCs (multinational companies) can be defined as companies that operate in more than one country with headquarters often in MEDCs and branch plants around the world, often with production focused in LEDCs.

| Advantages of TNCs | Disadvantages of TNCs |
|---|---|
| Provide employment and higher standards of living. | Jobs may be low-skilled or scarce due to mechanisation. |
| May improve the level of expertise of the local workforce and allow technology transfer. | Most profits are exported back home (HQ in MEDC). |
| Foreign currency is brought in, which may improve the balance of payments. | Priority may not be given to health and safety and pollution. |
| Can lead to the multiplier effect and widen the country's economic base. | May unduly influence political decisions. |
| | Manufactured goods are for export not the home market. |
| | Company may pull out at any time as decisions are made outside the country. |

*'Just in time' production is a Japanese method that minimises the costs of holding stocks of raw materials by ordering them just before they are required. Very efficient ordering and delivery systems are needed to maintain production. Component suppliers may locate near the main firm.*

## Use your knowledge

 **1** Why do big companies locate in the CBDs (central business districts) of large capital cities?

*Think of the advantages.*

 **2** What are the relationships between the percentage working in the secondary sector and the level of economic development of a country?

*Think of changes over time.*

 **3** What factors have caused a decline in manufacturing industry in the UK over the last 25 years?

*Think domestic and global.*

 **4** For a country you have studied, examine how inward investment has influenced the economic and social well-being of the population.

*Give a balanced answer.*

10 minutes

## Test your knowledge

1 Which is the largest employment sector in the UK?

2 Why is there increased demand for retail services?

3 What factors have increased the number of tourist visits worldwide?

4 What impacts does tourism have on geographical environments (give four categories of impact)?

5 Suggest a conflict that might exist in English national parks.

6 What does the term 'landscape value' mean?

✔ If you got them all right, skip to page 77

# Service Industries

## **Improve** your knowledge

**1** Service (tertiary) industries can be divided into three groups:

Key points from AS
in a Week
The sector model
page 80

- Consumer services offer services directly to the consumer and include retail, leisure and tourism.

- Producer services are those that help businesses carry out their functions.

- Public services are vital facilities that allow modern economies to operate.

MEDCs have a higher percentage of people employed in service industries; it is the growth sector in Britain. Services increasingly employ female, part-time workers in tourist service industries and producer services. Services have grown over the last 50 years as a result of:

- increased affluence and therefore demand for services

- changes in manufacturing that require more producer service workers

- globalisation of industries that requires more administration

- development of new technology, such as computers, that requires workers to program and service them

- reclassification of manufacturing jobs as producer services.

**2** Retailing can be defined as the sale of goods and services to the public; including high- and low-order goods. Retailing is often extended to include consumer, professional and financial services, such as hairdressing, solicitors' practices and banking. Food, drink and entertainment outlets may also be included.

Retail functions are traditionally located in the CBDs (central business districts) of urban areas with district centres containing lower order functions. In MEDCs there has been a shift of location to the urban-rural fringe. The comparative advantage of city-centre locations has been reduced due to increases in rents, lack of room for expansion, parking problems and security problems. Out-of-town locations offer easier access to suburban populations with free parking and large areas for extensive stores.

# Service Industries

**3** **Leisure and tourism** is the world's fastest growing industry and employs 7% of the world's population. Leisure activities can be enjoyed at home, whilst tourism involves visiting a place for at least one night, but less than a year.

Different destinations have been popular over time. From the 1920s, seaside resorts such as Blackpool became very popular; package holidays thrived in the 1960s and 1970s. Nowadays, an increasing number of tourists are seeking exotic destinations in LEDCs.

*It is difficult to differentiate leisure and tourism from other forms of employment. E.g. restaurants supply tourists and businesses.*

Rise and fall of selected tourist resorts over time

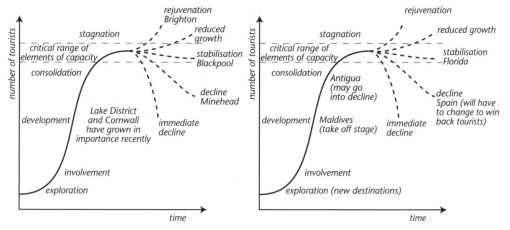

*Tourist visits to LEDCs have more than doubled over the last 20 years.*

**4** Tourism is blamed for massive environmental, cultural and social damage: polluted beaches, degraded coral reefs, displacement of local population, low financial return to the host country and abandonment of traditional economic activity.

**Eco-tourism** is an environmentally friendly alternative form of tourism. It is a niche market within the tourist industry, but is the fastest growth area. However, this type of tourist excursion is small-scale and expensive, and methods to make mass tourism more sustainable are required.

*Excursions to tropical rainforests can be small-scale and non-damaging.*

**5** Protection of the countryside in England and Wales:

- **National parks** are areas of outstanding scenery where human activity is carefully controlled and where the environment is preserved or improved for the enjoyment of people and for the conservation of native plants and animals. There are national parks in many parts of the world. In the UK people live and work in the parks and much of the land is owned privately, therefore the public do not have automatic access to the land.

- Environmentally sensitive areas are rural areas of national environmental significance worthy of preservation, e.g. the marshes of the Norfolk Broads.

- Designated Areas of Outstanding Natural Beauty cover 10% of the land area of England and Wales and are given special protection within the National Parks Act of 1949. They are smaller than national parks and include areas such as the North and South Downs (in southern England) and the Northumberland coast.

- Nature reserves protect endangered species.

**6** The location of service industries can cause conflict between those in favour and those against. There is much conflict within national parks regarding the best use of the land. There may be local opposition to opening a new out-of-town supermarket or a new airport runway. Different geographical environments are affected:

- the natural environment (air or water pollution, loss of habitats)

- the built environment (buildings and roads)

- the economic environment (effect on local jobs and prosperity)

- the social environment (effect on the community/new ways of life).

Public opinion changes, so issues that are important now, such as supermarkets on greenfield sites, were once not so. Decisions over the use of a landscape resource may be taken after an environmental impact assessment has taken place, along with other forms of consultation known as intervention planning, but large companies may have more financial power than other interested parties.

**40 minutes**

## Use your knowledge

**Hint**

**1** What are the typical features of retail parks?

*Think of layout and location.*

**2** Distinguish between primary and secondary resources for tourism and recreation.

**3** Discuss the statement that the economic benefits of tourism almost always outweigh the environmental costs.

*Give a balanced viewpoint.*

**15 minutes**

## Test your knowledge

1 What is the 'North–South divide', in a world context?

2 What is the most commonly used indicator of development?

3 What are the three most basic human needs?

4 Where are the core and periphery regions of Europe?

5 What types of appropriate technologies could help poor rural farmers?

6 What are the problems of large-scale prestige projects such as building dams?

7 What is the IMF?

8 What is a 'vicious cycle' of poverty?

9 How long do rural African women spend preparing food on average per day?

10 What could be the demographic impacts of AIDS on an African country?

## Answers

1 The continents of North America, Europe and northern Asia are rich and the continents of South America, Africa and southern Asia are poor. Australasia is an anomaly because it is rich, but is in the South – it is included in the rich North 2 GNP (Gross National Product) per capita 3 food, water and shelter 4 Core is the 'hot banana' between the English Midlands and northern Italy. Periphery is the Mediterranean region and Northern Scotland/Ireland. 5 simple wooden plough, trickle irrigation, biogas converter to power a low wattage electric cooker, mechanical rice huller, etc. 6 very expensive, do not benefit the poorest people (often they lose their land) and have serious environmental side effects 7 International Monetary Fund 8 when poor people cannot break out of the poverty trap because lack of income, poor health and education prevent any improvements to lift them from the cycle 9 up to 5 hours 10 loss of economically active workforce to provide for young and old dependants and to provide a tax base to run the country's public services

✔ **If you got them all right, skip to page 85**

# Development Issues

**45 minutes**

## Improve your knowledge

**1** There are variations in the economic development of different countries within the world. The Brandt report in 1980 distinguished the rich North from the poor South. The way countries are classified disguises marked variations in development and standards of living both within and between countries – even those that are nominally within the same category.

What does seem to be a continuing trend, however, is that the gap between the rich and poor countries is widening. Many see this as a result of the policies pursued by richer countries. The most telling statistical measure of this is that the rich North has 25% of the world's population but 80% of the world's income whilst the poor South has 75% of the population and 20% of the income.

**Key points from AS in a Week**

Rostow's model of economic development
                    page 79

Strategies to reduce urban primacy in LEDCs
                    pages 74–75

*The share of the poorest fifth of the world's population in global income has dropped from 2.3% to 1.4% over the last 10 years.*

**2** Indicators of development include:

- Gross National Product (GNP) per capita is all the monetary value of goods and services within a country (both visible and invisible) divided by the total population. GNP is an unsatisfactory measure for countries without a market economy and with a high level of subsistence activity. GNP per capita hides inequalities within countries and gives little indication of social development.

- Balance of trade is the balance between imports and exports of visible goods. Balance of payments includes invisible services such as finance and tourism. LEDCs tend to have a negative balance of trade and payment, with exports of raw materials and foodstuffs and imports of manufactured goods.

- The Human Development Index (HDI) gives every country a score from 0 to 1 based on its citizens' life expectancy at birth; education (adult literacy rate and number of years' schooling); and income (converted to purchasing power parity – what an actual income will buy in a country). However, the HDI does not measure human rights, but it does allow countries that have a better or worse level of well-being than their GNP per capita may suggest, to be identified.

- Population characteristics are likely to indicate the development of a country; they include birth rates, fertility rates, infant mortality, death rates, etc.

*Britain has enjoyed a surplus of invisibles which has helped pay for the long-standing deficit on visible trade.*

*Canada has the highest HDI figure of 0.99. Sierra Leone is at the bottom of the list with 0.41.*

- Employment structure: LEDCs are likely to have a high proportion of people working in agriculture at a subsistence level and in the informal sector.

- Rural-urban spread: LEDCs are likely to have a higher proportion of people living in rural areas.

- The position of women and children may be poor in LEDCs, with women denied equal access to education, work, inheritance, etc., and children exploited as cheap labour.

- Daily calorie consumption: many people in LEDCs are malnourished and eat less than 2200 calories per day due to an uneven distribution of food and the high cost of food relative to income.

*In Ethiopia 87% of the population live in rural areas and only 4% have sanitation facilities.*

**3** Development is the use of resources and application of available technology to bring about an increase in the standard of living within a country. Earlier views on development emphasised economic expansion and increased output, but the current view is much broader, involving cultural and social advancement as well as technological change and economic growth. Distribution of income and an appreciation of human needs – the basic provision of food, housing and water for everyone – are seen as fundamental.

**4** There are a number of theories that attempt to explain why some countries are more developed. Rostow's theory has been dealt with in *AS in a Week*. Bark and O'Hare (1984) claim that although MEDCs have moved through Rostow's five stages, it is unlikely that many LEDCs will follow the same pattern. They suggest that a dual economy may exist (one or two developed cores surrounded by a poorly developed periphery) and that many LEDCs will never make the transition from a traditional agricultural economy to an advanced industrial one. Alternative development strategies and ways of measuring development need to be pursued.

Myrdal's theory of cumulative causation:

- Growth poles are areas that are successful and therefore attract even more economic activity, labour and investment.

- The process of circular and cumulative causation is a result of the multiplier effect. Expanding economic activity in an area creates extra employment and raises the total purchasing power of the population.

- The population has a greater disposable income, so linked industries are attracted.

- The larger labour pool, expanded local market and improved infrastructure, attract even more economic activities that require a larger threshold population.

- Industries benefit from the agglomeration economies that a larger area allows (see the chapter on Service Industries).

- The process of cumulative growth becomes self-sustaining, fuelled by migrants, potential entrepreneurs and capital attracted from less advanced surrounding areas.

- The spatial concentration of resources towards the core causes backwash effects on the surrounding area (the periphery).

- Cumulative causation permits upward spirals in the core and downward spirals in the periphery.

- Peripheral areas gradually develop later through spread from the core, because the expansion of the core increases demand for food and resources from the periphery.

- Innovations may diffuse from the core, and decentralisation from the core stimulates growth in the periphery as land prices and congestion become too high in the core.

- However, the gap between the core and the periphery widens if backwash is greater than spread. This would be typical of a downward spiral, whereby investment and skilled people are attracted to the core.

*Backwash effects occur when resources and people are attracted to the core – this drains the periphery.*

*Hirschman termed the spread effect from the core 'trickledown'.*

Friedmann's theory argues that development of a country is dependent on the spread of wealth from the core to the periphery.

Friedmann's core and periphery structure

**Stage 1**

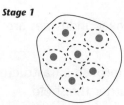

Relatively independent centres, no hierarchy.
Each town lies at centre of a small region.

**Stage 2**

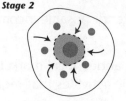

A single strong core with a periphery.
Labour and capital move to the core.

**Stage 3**

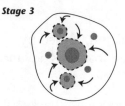

A single, national core with a number of
peripheral sub-cores.

**Stage 4**

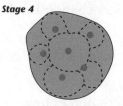

An interdependent system of cities with a
maximum growth potential for the whole
country.

**5** In many rural regions of LEDCs, a vicious cycle of poverty exists within the economy. Farmers lack money and cannot invest in better seeds, equipment or fertilisers. This inevitably results in poor crop yields, giving them low income that in turn results in a lack of investment in the system. Loans and grants from the government or international funds will help to break the cycle as long as the farmers are also educated in more efficient agricultural methods. This type of development is most effective if approached on a small scale, using appropriate technologies and from the grass roots (i.e. directly aimed at the recipients). Micro-credit unions can help farmers borrow money without becoming heavily in debt.

*Grass roots strategies are often carried out by organisations such as Oxfam and are called 'bottom up' approaches to development.*

**6** Aid is the giving of resources by a donor country or organisation to a recipient country with the main aim of improving the economy and/or the standard of living of people within the recipient country. The resources may be in the form of money, food, goods, technology, training or skilled people. There are three main types:

- Bilateral aid (from one country to another) is often tied – the donor country imposes conditions such as contracts for building or preferential trade links.

- Multilateral aid is when richer countries give money to an organisation such as the World Bank or the EU which then distributes the money to poorer countries. Aid may be withheld if they disagree with the recipient country's economic and/or political system.

- Voluntary organisations such as Oxfam or Comic Relief collect money from the general public and spend it on specific, usually small-scale, projects or disaster relief.

*Voluntary organisations are called non-governmental organisations – NGOs.*

**7** Some of the most effective aid can be delivered through alternative or intermediate technology, which is used in LEDCs where hi-tech industries are expensive or inappropriate for the needs of people and the environment. Projects should be labour intensive, use local skills, local technologies and resources and be low cost. They also need to be sustainable and in harmony with the local environment. To improve food security it is more appropriate to improve existing farming methods and equipment than to attempt to introduce totally different production methods (a disadvantage of the green revolution). A criticism of food aid is that it undermines local production.

*NGOs like The Intermediate Technology Group work closely with women to produce effective labour-saving technologies such as corn peelers and energy-efficient stoves.*

**8** The role of women and children must not be forgotten when delivering aid strategies. Women are usually responsible for food production and preparation, child rearing and domestic tasks, particularly in rural areas. Children from as young as three perform many tasks that provide income for poor families. The most marginalised people are often minority groups, rural dwellers, disabled people and women. In LEDCs there are many more illiterate women than men, and women's wages are lower than those of their male counterparts.

**9** Many LEDCs are in considerable debt as a result of loans accumulated through aid programmes in the 1970s. The lending agencies granted long repayment periods in anticipation of a good return as economies started to expand in LEDCs. The recession in industrial countries in the 1980s led to a fall in trade and commodity prices and an increase in interest rates. Many countries were using new loans and aid to pay interest payments on previous debts. The debt crisis became public in 1982 when Mexico announced that it could not pay its foreign debt. Solutions to the debt crisis include:

*Informal employment has increased sharply in LEDCs, offering low-income temporary jobs instead of well-paid reliable employment.*

- Structural adjustment policies implemented by the IMF. This process of economic change tries to integrate indebted countries into the free market of the global economy, but demands decreased public spending, which often hits the poor hardest as education and health services are reduced.

- Fairer trade, because LEDCs are at the mercy of fluctuating primary commodity prices on the world market. Globalisation has created such a degree of interdependence that a change or decision in one place can affect the operation of industry elsewhere. Trade barriers in the form of tariffs and quotas stop LEDCs penetrating the MEDC market (see the chapter on Manufacturing Industries).

- Cancellation of debt.

- New forms of credit such as micro-loans given at the 'grass roots' and directed to people who need it most.

The debt situation has eased in some countries and some debt has been written off. However, it remains a serious obstacle to development.

**10** The epidemiological transition shows how, as countries develop, their populations go through different stages of diseases and causes of death. It helps planners project future health care needs both nationally and internationally, and draws attention to the problem of the developed world's ageing population. One of the world's most pressing health problems is AIDS, which is widespread in many African and Asian countries, and will have a huge demographic and therefore economic impact, presenting more barriers to development.

*Large TNCs wield such economic influence, and many governments have become locked into a 'trade, aid, debt for development' scenario, so they find it difficult to make progress in terms of human development.*

# Development Issues

## Use your knowledge

40 minutes

Hint

**1** Discuss the pros and cons of using GNP as the sole indicator of development.

*Give balanced views.*

**2** What are the arguments for and against giving aid?

**3** What other development strategies are options for LEDCs other than a reliance on manufacturing industries?

*Think of primary and tertiary industries.*

# Synoptic Assessment

The synoptic assessment part of your examination is one which many students dislike. It is seen as being difficult to score highly on, but with practice (using other synoptic questions and past papers), you will be able to tackle such questions with confidence.

This type of question sometimes includes a decision-making exercise (DME) and it gives you the opportunity to demonstrate your knowledge and understanding of:

- a range of different geography subject areas
- connections between these different subject areas
- the importance of the human impact in various themes and issues.

You are NOT expected to write long explanations of the processes and theories you have studied in physical geography (e.g. glacial erosion and weathering) and human geography (e.g. models of urban land use).

Instead, you should apply this knowledge to the locality in the question. Do not start writing a description of the coastal protection measures in East Yorkshire when the questions are all based on the Dorset coast!

A good answer to this type of question should:

- always be relevant to the theme of the question (perhaps water supply in East Anglia)
- show understanding by using the appropriate terminology
- describe how the natural and human environments are affecting the issue
- have an ordered, logical and step-by-step approach to the answer.

Most issues will involve the interaction between the natural environment and human activity, perhaps a proposed new quarry or road in a national park, various ways of tackling coastal erosion in a particular locality or the planning issues surrounding a new retail park.

Some exam boards include an Ordnance Survey map which you should use in your answers. You could use the map to work out accurate distances or to identify features on a photograph.

Some examination boards send out pre-release material, which your teacher will give you a copy of. This usually consists of a few pages of text, graphs and photos about a geographical issue. In the two or three weeks before the exam your teacher will set classwork using this booklet so that you will become familiar with the issue well before the exam. This preparation is essential, as the questions will assume you are already fairly well-informed on the topic.

*If your exam board doesn't issue pre-release material, you should study the exam booklet for at least a full 10 minutes before you begin to write.*

## How to use the pre-release material

- Read the booklet repeatedly to fully understand its content and layout.

- Use maps/atlases to study the locality/surrounding geographical area.

- Look up any unfamiliar words or phrases in your textbook.

- Revise any unfamiliar theories and processes using your textbook or class notes (if the issue is coastal protection, for example, make sure you know about the types, causes and effects of coastal erosion; if it is flooding then revise all about hydrology and hydrographs).

- Try to consider possible questions which might be asked.

Sometimes you will be expected to assume the role of a particular person (e.g. the farmer opposed to the new quarry) or a particular organisation (e.g. the environmental group opposed to the new quarry). Different people and organisations will have different reasons to support or oppose the changes or developments. Environmental impacts are sometimes accepted, as the benefits of job and wealth creation would be great.

As you write your answers, you will need to list, and then describe, the different factors involved, such as:

- long-term as well as short-term effects of the proposal or issue

- local and national considerations as well as effects on the immediate locality

- the advantages and disadvantages of the proposal or issue.

Some people find it easier to structure their answer if they divide the impacts into:

- social and cultural impact – on people and their lives

- economic impact – on jobs and local business

- environmental impact – on ecosystems and landscapes.

## Some common themes used for the synoptic assessment

- estuaries: land-use, influence of the source river, impact of 'barrage' construction

- growth of tourism in a scenic area – impact on the human and physical environment, conservation and management

- town morphology – retail catchment, reasons for and impact of out-of-town retail development

- study of a forest in the UK – maps and graphs showing vegetation composition, influence of physical and human factors on the vegetation, impact of 'forest tourism'

- coastal ecosystems and management of coastlines.

Not all locations will be in the UK; you may have maps and data from anywhere in the world. However, the main theme remains the interaction of the human and physical environment.

(a) Outline two types of evidence that can show the scale of an earthquake hazard in any one place. (2 marks)

(b) Describe the effects of the hazard. (4 marks)

(c) List the ways in which the hazard can be predicted and managed. (4 marks)

The diagram below shows the nutrient cycle in the northern coniferous forest.

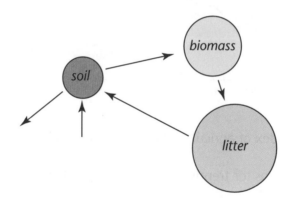

(a) Why is the litter layer the largest nutrient store? (3 marks)

(b) Using a similar diagram with labels, explain why the nutrient cycle in the tropical rainforest is different to the one in the diagram. (4 marks)

(c) With reference to the tropical rainforest, describe three possible effects of deforestation on the soil. (3 marks)

**3** Study the graph below, which shows trends in agricultural production.

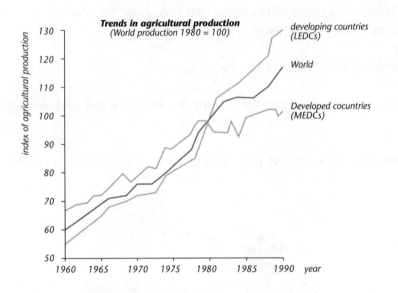

**Trends in agricultural production**
(World production 1980 = 100)

- developing countries (LEDCs)
- World
- Developed cocuntries (MEDCs)

*index of agricultural production*

(a) Why is an index of agricultural production used? (2 marks)

(b) Suggest reasons for trends in agricultural production between 1960 and 1980. (4 marks)

(c) Suggest reasons for trends in agricultural production between 1980 and 1990. (4 marks)

**4** Study the map opposite, which is a simplified version of the core–periphery structure in Europe.

(a) Define the term 'core'. (2 marks)

(b) Explain the location of the Western European core. (2 marks)

(c) Suggest why the core is moving. (3 marks)

(d) What factors may have encouraged the growth of the 'high-tech' zone? (3 marks)

Core and periphery

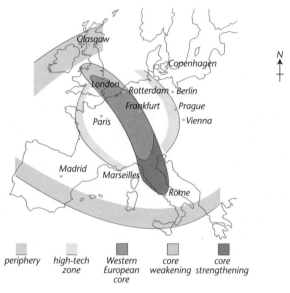

periphery | high-tech zone | Western European core | core weakening | core strengthening

## Essay Questions

Answer two of the following questions.

 Explain the different factors that cause the type and distribution of deposits to vary on a beach. (20 marks)

 Using examples, examine the impact of different human activity in mountainous glaciated environments. (20 marks)

 Describe the causes and consequences of soil structure alteration due to human activity. (20 marks)

 Examine the extent to which the physical environment can be managed sustainably. (20 marks)

 How do governments attempt to influence the location of manufacturing industry? (20 marks)

Discuss how the increasing trend towards the globalisation of production has influenced one LEDC that you have studied. (20 marks)

# Use your Knowledge Answers

## Plate tectonics and hazards

**1** (a) This invites an explanation of the San Andreas Fault and of conservative plate margins. You should fully explain what happens at a conservative margin, stating that earthquakes are a result of a sudden release of stress on rocks under pressure along the margin or fault line. Plates should be named and the relative rates of movement given. Explain that volcanic activity is less likely here due to the absence of subduction and the associated melting of rock.

(b) Use terminology where appropriate and describe each 'method' carefully, rather than just listing lots of different methods. Refer to actual places where such measures have been taken. You could mention the contrast in ability to adjust between LEDCs and MEDCs. Include forecasting and prediction, modifying the hazard and damage limitation measures.

**2** (a) Describe processes in detail. The question refers to 'hotspot' areas – areas of stationary crust over intense convection currents. Crust is melted from below, is weakened and starts to collapse. This causes fractures allowing volcanoes to develop. If the plate slowly moves, volcanoes become extinct and new ones develop in their place. Example: Hawaii.

(b) Define both terms. Mention how magnitude may be measured – Richter scale for earthquakes. Try to refer to hazards from different categories, perhaps a tectonic and a climatic hazard. Impacts should be described in terms of the effect on the economy, on the environment and on people's lives. Case studies should address the role of frequency and magnitude at the locations concerned, such as studying cyclone tracks or analysing patterns of volcanic eruption. You should include risk assessment and preparedness. Magnitude affects the degree of destruction but the characteristics of the area and its level of development are other important factors. High magnitude cyclones may have less impact as prediction and preparedness is possible days in advance.

## Meteorology and climatology

**1** (a) Temperature falls away from the centre of the city, changing from 10.6°C to 5°C. You could calculate that the rate of change is about 0.4°C per km (change of 5.6°C over 13–14km). Temperatures are lower close to rivers by up to 2°C, except in the city centre, where the river seems to have no effect.

(b) Due to heat from buildings, vehicles, domestic heating and industry. Many buildings store heat up (they may be designed to do so) and release the heat slowly at night. Dust and dirt in the air over a city can insulate against heat loss to the air. Lower urban wind speeds prevent the winds from blowing out as much warm air.

(c) Urban areas receive more precipitation due to greater convection due to greater warming; dust provides condensation nuclei for formation of raindrops; more water vapour in the atmosphere due to water usage and steam emissions from industry. There is more snowfall over rural areas due to lower temperatures.

**2** (a) You should fully describe the natural greenhouse effect – why it is needed for life on Earth. Refer to the various energy flows due to insolation, terrestrial re-radiation, etc. Then go on to explain how humans have accelerated the process, mainly by use of fossil fuels, but also refer to new evidence such as methane production from large-scale cattle ranching, etc.

(b) Official responses by governments have arisen due to international agreements entered into at various environment summits such as Rio de Janeiro, Kyoto and most recently, the Hague in November 2000. Discuss how any agreement a country makes tends to accommodate its need for economic development – China is unwilling to reduce coal usage as it is the basis for its economic growth. Countries only agree to reductions that are unlikely to affect them.

## Coastal environments

**1** (a) Describe the alternating bands of different rocks. The clay/sand is softer and so has eroded more than the others. The bands of rock run perpendicular to the coast and this produces a bay and headland sequence or discordant coastline.

(b) The groynes – describe what they are – are evidence that longshore drift is an issue on this coast. Describe the longshore drift process and state that here it takes place in a west to east direction. Explain how the groynes trap sand and shingle and widen the beach. Mention that the town must be protected – possible tourist industry.

(c) Lots of possible lines of discussion here. Mention trampling of vegetation and all its consequences for erosion and habitat destruction. Litter could be a problem, campfires and so forth. Try to refer to coastal species (marram grass, sand couch, nesting birds) which might be affected. Visual pollution (tents), noise pollution and pollution from vehicle fumes may occur. The need for a new road to the town may cause habitat loss.

**2** You should concentrate on the marine processes of erosion and deposition but also evaluate the importance of other factors which have influenced your chosen coastline, such as relief, sub-aerial processes of weathering and mass movements, rock type, longshore drift and the various human modifications which may have been made, such as sea walls. You need to carefully learn the names and formation of the features of a particular coastline. Depending on your case study, you may describe how a beach profile has been caused or how the marine processes of corrasion and quarrying can produce stacks, caves and blowholes. If relevant, the landforms created by eustatic or isostatic sea level changes could be described. Include a sketch map of the area which locates the features you are describing.

# Use your Knowledge Answers

## Glacial environments

**1** (a) Describe the four processes: basal flow (slippage), extending and compressing flow, creep and occasional surges. The base is lubricated by water which has melted due to pressure from the mass of the glacier above. Go into lots of detail when describing the processes.

(b) The graphs show that the temperate glacier, such as those in the Alps, has faster ice flow, especially in the upper layers near the ice surface. Flow at the base is again faster in the temperate glacier. In the polar glacier, there is no flow at all at its base. In both cases, flow is slower at the edge of the glacier, where it meets the valley sides.

(c) In the temperate glacier, the temperature of the ice is just below 0°C so it melts more readily at its base, providing lubrication. The process of 'creep', when the ice particles move forward then refreeze, should also be explained. Movement at the sides is slow due to friction from the valley walls. The ice may freeze around valley protrusions, slowing its progress.

**2** (a) In theory, you can get full marks using a fully labelled diagram, but this is risky. Use a well-labelled diagram but expand on each label in a separate paragraph below. Describe size, shape, surface and orientation. Explain in terms of the action of the meltwater streams in transporting and sorting sand and gravel. Refer to seasonal fluctuations in discharge and velocity and their effects (if relevant to your chosen landform).

(b) Some reference can be made to the shape of landforms indicating their formation; eskers can meander in the same way as their parent streams; terminal moraine seems to have been 'bulldozed' by the ice. Main reference is to the unstratified, angular and unsorted ice-borne deposits and sorted, stratified (layered) and rounded fluvioglacial deposits.

## Periglacial environments

**1** (a) Mean annual temperature varies from –12°C near the poles to –1°C in lower latitudes. In the far North, winters may fall to –50°C and summers are very short, but in areas with only sporadic permafrost, summers are longer and the temperature may reach 2 or 3°C. Precipitation, as snow, is generally low at around 300mm p.a.

(b) Rock weathering: freeze-thaw needs fluctuations above/below 0°C, so that water thaws and trickles deeper into the rock before re-freezing. Most active in April–June, late spring and August–September (autumn) because there is no fluctuation in winter. Mass movements: solifluction needs the active layer to thaw so occurs in spring and summer, (May–September) but is at its maximum in May when the active layer is still thin and so moves easily. Fluvial (river) processes often stop altogether in winter. Erosion and transportation are

greatest during the spring thaw when discharge is very high. Deposition occurs more in August–September when the freeze-up begins and discharge drops.

(c) Rivers have very volatile seasonal discharge, with the maximum in spring/early summer. The river can carry much rock debris (from freeze-thaw) and gravel but when discharge falls quickly at the end of summer, it is deposited in 'barrows' which divide the channel.

**2** Refer to cold mountainous regions as well. Describe the problems of the natural environment: cold; lack of available water; near impossibility of agriculture; poor soils; permafrost; long-term snow cover. Expand on each point. Describe and briefly evaluate attempts to settle in such areas: new infrastructure and improved accessibility; successful establishment of tourist resorts in difficult areas (such as Alpine ski resorts); resources such as oil have been successfully exploited in difficult areas; government services have improved some aspects of the lives of indigenous inhabitants; refer to application of technology such as 'utilidors' and techniques for constructing buildings. Refer also to the drawbacks – various impacts on the indigenous peoples, environmental pollution and other damage from ski resorts, oil industry; impact of buildings and roads on the permafrost. You could argue that it will always be difficult to settle there due to political opposition.

## Ecosystems and fragile environments

**1** (a) Refer to the general reasons for clearances and alterations. Rapid global population increase in 20th century, areas with most rapid increases often have most widespread deforestation/clearance. Economic development causes clearance for land for industry and housing. In the early stages of economic growth, timber may be used for fuel and construction in vast quantities. Political decisions to open up new areas, such as Brazilian interior. MEDCs (such as UK) have long history of economic and population growth and so little CCV (mixed deciduous forest) remains. In LEDCs, vast clearances and changes occurring as these countries rapidly develop (Brazilian or Malaysian rainforest), but still many undeveloped areas still with CCV. Refer to the establishment of protected areas and national parks. These may not be enforced effectively in some countries. Good answers should finish by drawing a brief conclusion to the points you have described.

(b) There is no specific request for case studies but you will score more highly if you base your answer around a particular region – perhaps the Sahel. You should describe the human activities which cause land degradation and soil erosion. Include definitions where appropriate. You could refer to salination of soil, subsistence agriculture, nomadic herders and irrigation. You should describe the ecosystem and say how it is affected by human activities. Try to refer to flora and fauna. Your case study will provide the framework for the physical processes (excess run-off, wind deflation) and concepts such as desertification, albedo and changes in nutrient cycling.

**2** (a) Flow A = leaf fall/addition of organic matter
Store B = litter layer

(b) Fungi and bacteria break down the leaf litter and other organic matter into humus by the process of humification. Earthworms and other soil fauna incorporate the humus into the soil.

(c) Nutrients are taken up into the biomass (plants) in solution as the nutrients are dissolved in the water taken by the plants' roots.

(d) More surface run-off so more nutrient loss on the surface. Greater leaching of nutrients down through the soil horizons. More water available for plants so more water (and nutrient) uptake for growth. More input of mineral nutrients from weathering of the parent rock, chemical weathering is increased by plentiful precipitation.

## Population and resources

**1** (a) Bruntland Commission (1987) definition: 'development that meets the needs of the present without compromising the ability of future generations to meet their own needs'.

(b) Pressure on resources, resource depletion/exhaustion (shortages, famine and Malthusian checks); environmental pressure and degradation (Club of Rome realised); conflict within and between societies over resource ownership and right; potential for more migration flows between LEDCs and to MEDCs.

(c) Changes in consumption patterns, political and attitudinal changes, costs of switching to renewable resources and recycling, distributing resources more equally. Role of MEDCs is crucial in aiding the process in LEDCs. Stress the importance of Earth Summits and global agreements.

**2** Accessibility, quality and quantity of the resource. Technology – due to the uneven nature of the sea bed and rough seas, new designs of oil rig were required to exploit new resources. Economic viability – the OPEC oil crisis made extraction of the oil viable. Exhaustion of other energy resources – the most accessible sources of coal had been exploited and costs were rising. Political will and multi-national interests. Rising demand for an efficient high energy fuel for domestic, industrial and transportational purposes.

## Rural areas and food supply

**1** Loss of habitats, loss of biodiversity, pollution of surface and ground water from effluent and nitrates, soil erosion, introduction of diseases/chemicals into the food chain,

salinisation; explain these in relation to intensive agricultural methods.

**2** Advantages: may be a commodity that fetches a high price on the world market and guarantees income; may be highly labour intensive; may attract inward investment and therefore modern farming practices. Disadvantages: vulnerable to fluctuations in world prices and demand; crop may fail and there is very little else to export; primary commodities are of low value compared to manufactured goods, therefore balance of payments is negative; best quality land is used to grow the crop and not food for the domestic market; often companies are controlled by TNCs or former colonial administrations so profits leave the country.

**3** Large farmers have succeeded because they had money to invest in HYVs, irrigation, mechanisation, fertilisers and pesticides. Some small farmers have increased output, especially if land reform has been part of the package or if they have received grants. However, many have become landless labourers. General increase in food output has benefited people.

## Manufacturing industry

**1** Prestige of capital city, advantages of concentration (agglomeration/linkages), inertia, nodal point for communications, more skilled labour market, higher demand for product.

**2** A high percentage in primary and a low percentage in secondary is a good indicator of low level of development. Secondary sector increases as development occurs (at the expense of the primary sector). With further economic development, the relative percentage in secondary is reduced by an increase in the tertiary sector. Change occurs as the needs of the economy change.

**3** Domestic: mechanisation, rationalisation, decline of the heavy industries such as iron and steel, car manufacture, etc. Global (of course global issues can be the catalyst for domestic changes): competition from imports, TNCs moving production to cheaper locations in LEDCs, world recession, loss of comparative advantage.

**4** Use a detailed case study to illustrate this question; show the examiner that you know facts and figures. You could use a range of examples (e.g. Japanese investment in Britain or inward investment in NICs). Give a balanced answer between economic (balance of trade, prosperity, job creation) and social (increased standards of living, training, improved infrastructure, etc.) but remember the effects on people can be positive and negative.

## Service industries

**1** More peripheral edge-of-town locations, often good road/motorway access, relatively new construction and design which pays attention to detail, emphasis on single-storey open-floor plans, large space allocation to vehicle movement and parking.

**2** Primary: the original features that attract a tourist to visit (scenic amenity, climate). Secondary: those facilities that are provided to serve the needs of tourists (accommodation, catering, entertainment). In the case of a location like Blackpool, the secondary facilities become the prime attraction in themselves.

**3** Economic: jobs, multiplier effect of greater income, more developed infrastructure, greater income in existing services, more investment in facilities (you could include social/cultural issues too). Environmental: congestion, pollution of air/soil/water, new buildings damage habitats, waste disposal issues, resources exploited to supply tourists. You must evaluate whether economic outweighs environmental by examining the factors through a variety of case studies.

## Development issues

**1** Gives a universal indicator of wealth which tends to suggest economic development, but masks any form of social development related to health, education, etc., also does not indicate the spread of wealth between people within a country.

**2** For: gives immediate disaster relief, helps implement projects, improves education which may increase agricultural output and employment and decrease birth rates, improves health and living conditions. Against: often used to exert economic and political pressure and may be tied, often does not reach the poorest people, may fund inappropriate prestige projects, may undermine local activities, may increase debts.

**3** Alternative development strategies could be through agriculture (agribusiness and sustainable farming methods), tourism, fairer trade (regulating trade restrictions) or aid (small-scale appropriate projects).

## Answers to mock examination

**1** (a) You could consider as evidence: numbers killed or injured (over the long term as well as secondary effects), numbers made homeless, degree of damage to the built environment, extent of damage to crops and livestock and the degree of disruption of harvests in the succeeding years, the number of or total value of insurance claims. Comment on the differences between MEDCs and LEDCs and how some types of evidence (insurance) may be inappropriate.

(b) Ground movement/tremors, landslides, mudflows, avalanches, rockfalls, tsunamis, 'ground liquefaction', collapse of man-made structures (dams and buildings), damage to infrastructure. Refer to secondary effects as well: homelessness, disease and economic/agricultural collapse. Link secondary to primary effects – disease caused by contaminated water, caused by broken sewer pipes.

(c) Fault zones can be monitored by scientific equipment; faults can be 'lubricated' with vast quantities of sand; adaptations to buildings and other preparations. Describe how schemes differ in LEDCs due to lack of funding.

**2** (a) Low temperatures and acidic litter (pine needles) means earthworms and other incorporators work very slowly and so litter builds up.

(b) Your diagram should have a small soil store, very small litter store, large biomass store and much broader arrows. The soil store is lower due to rapid take-up by dense vegetation, and leaching due to high rainfall. Rapid transpiration loss encourages vegetation to take up large amounts of soil moisture, containing nutrients. Large biomass due to constant growth, high temperature and rainfall so rapid photosynthesis. Soil fauna/incorporators work under ideal conditions so litter store is small.

(c) Reduced standing biomass so reduced input of humus from the litter. Reduced humus leads to breakdown of the clay–humus complex and loss of nutrients. Soil structure is changed and soil may become easily erodable by wind or water. More surface run-off encourages loss of topsoil.

**3** (a) Index allows direct comparison of change and permits the use of different measurements (but it does not show absolute changes).

(b) MEDCs: intensification, massive use of fertilisers, enhanced breeding, subsidies, drive to self-sufficiency post-war (CAP), increased efficiency through specialisation and mechanisation. LEDCs: improvements in subsistence farming, green revolution, agribusiness for export.

(c) MEDCs: overproduction and surpluses lead to new controls such as quotas, set aside. LEDCs: globalisation of agricultural production, new commercial developments aimed at global market, inward investment from World Bank and TNCs.

**4** (a) An area of concentrated economic activity which is the focal point for a wider region.

(b) Fixed by the long established economic and financial dominance of London and the southeast through the Low Countries to the Rhone Valley, including the industrial concentrations in West Germany in Bonn/Frankfurt.

(c) Reflects the changing base of the economies. The traditional industry of the Midlands and northeast of England has declined. There has been an increase in lighter industry in the north of Italy.

(d) To the west: counter-urbanisation of small industry to more pleasant and less congested greenfield sites with skilled workforce, national government grants/subsidies to encourage growth away from the southeast, improved motorway links (e.g. M4). To the east: large, cheaper workforce released by the collapse of the eastern bloc countries and market opportunities in those countries.

**5** Larger material (shingle) nearer to the back of the beach, sand at the shore. The height of any berms indicates the strength of storms. Material is determined largely by local geology. The beach should be described as a system with inputs, stores and outputs (e.g. longshore drift). You could refer to human activity altering the system, such as groynes trapping sand. Try to use a case study of a beach you have studied and use terminology for processes and features.

**6** You could write about the extent and impact of quarrying or mining, dam construction for water storage and/or electricity generation, electricity transmission cables as visual pollution. Hill farming such as sheep farming in Cumbria and commercial forestry have their own implications for the environment and landscape. Tourism may result in unsightly buildings, footpath erosion or deforestation for winter sports. Consistently support your statements by accurately describing what has taken place in particular localities.

**7** Define and explain soil structure and point out its importance to plant growth and erodability. Compaction by agricultural machinery or recreational vehicles reduces root penetration, number of seedlings emerging and amount of $O_2$ and $CO_2$ exchange. Reduced 'infiltration capacity' increases run-off and soil erosion. Recreation such as motorcross can form gullies and reduce IC by up to 80%. Grazing and overgrazing cause compaction by animals' hooves. Refer to examples from specific localities, those in your own area can be just as relevant. Explain the processes fully; explain $O_2$ and $CO_2$ exchange carefully. It is valid to refer to soil structure improvement by farmers adding clay to sandy soil ('marling'), or deep ploughing to break up horizons.

**8** Use examples from a range of scales and a range of countries at different levels of development. Physical environments could include land or sea. You need to get across the idea that resources can be utilised within the concept of sustainability, but explain the factors that cause them to be exploited. By using case studies such as tropical forest resources of Indonesia, fishing in the North Sea, or the national parks of England and Wales, you can examine factors such as demand, need to make a profit, management, level of development, political will, etc.

**9** Use a range of examples to illustrate the various ways that governments and government agencies can influence the location of industry by creating comparative advantage. Methods include: regional aid, grants and tax exemptions, investment in infrastructure in a declining region, planning to steer industry away from core regions, attracting TNCs (or not). These can be delivered via international agencies such as the EU or via national government strategies such as urban development corporations and enterprise zones in the UK. You could use examples such as South Wales, Rhonda valley (EU funding), London Docklands (UDC), Italian Mezzogiorno and the NICs where government intervention has been paramount in encouraging manufacturing industry.

**10** To get top marks you need to discuss a wide range of effects of globalisation on your chosen LEDC, both positive and negative. These include the impact on local economies, impact on the environments, social effects, transfer of technology and training, changes to industrial output, etc.